HYDRANGEA PRODUCTION

GROWERS HANDBOOK SERIES
VOLUME 3

Allan M. Armitage, General Editor

HYDRANGEA PRODUCTION

Douglas A. Bailey

Department of Plant Sciences
University of Arizona
Tucson, AZ 85721 U.S.A.

TIMBER PRESS
Portland, Oregon

Cover illustrations

Fig. 4
Hydrangea distortion symptoms during high summer temperatures.

Fig. 5
Iron deficiency chlorosis.

Fig. 6
Left to right: 'Rose Supreme', 'Merritt's Supreme', 'Blau Donau', 'Dr. Steiniger', and 'Rosa Rita' produced using mini-pot techniques. Slide taken 10 weeks after first GA_3 spray and start of night break.

Fig. 7
A potential new florists' hydrangea: *H. macrophylla* subsp. *macrophylla* var. *normalis* 'Tricolor' grown using mini-pot techniques.

© Timber Press, Inc., 1989

ISBN 0-88192-143-2
Printed in Hong Kong

TIMBER PRESS, INC.
9999 S.W. Wilshire
Portland, Oregon 97225

Library of Congress Cataloging-in-Publication Data

Bailey, Douglas A., 1958–
  Hydrangea production / Douglas A. Bailey.
      p.   cm. -- (Growers handbook series ; v. 3)
  Bibliography: p.
  ISBN 0-88192-143-2
  1. Hydrangea macrophylla--Handbooks, manuals, etc.   I. Title.
II. Series.
SB413.H93B35   1989
635.9'3397--dc19   89-30774
                                                                 CIP

Contents

Dedication:

To Thomas C. Weiler and P. Allen Hammer,
for their endless support, guidance,
and contributions in Floriculture

Acknowledgments

It would be impossible to write a book on hydrangea production without the valuable materials borrowed from other sources. These include researchers as well as producers. I thank those whose research and experiences made this book a reality, full of information reported over a 60-year span.

I am grateful to a number of colleagues who read the manuscript for this book and offered valuable comments. Particularly, I am indebted and grateful to Allan Armitage for his time and effort in editing the multitude of manuscript revisions prior to final copy.

Introduction

The adaptation of woody plant species to greenhouse production has contributed many popular and novel crops to commercial floriculture. One such crop is the florists' hydrangea, or as it is known internationally, the hortensia.

Florists' hydrangea, *Hydrangea macrophylla* subsp. *macrophylla* var. *macrophylla* (Thunb.) Ser. is a member of Saxifragaceae (37,41). The hydrangea bears large (4–8 in. diam.), spherical cymose inflorescences composed of sterile flowers (florets) with enlarged, colored sepals on top and small, inconspicuous, fertile flowers buried below. The true petals are small, adding little to the flowers. Sepal colors include various tones of pink, blue, and white. The dark green foliage is 4–6 in. long and 3–4 in. wide and provides a handsome accent for the colorful flowers.

Hydrangeas have been an important greenhouse crop for many years. They are forced into flower from Valentine's Day through Memorial Day and are an important Easter and Mother's Day crop. Census figures, however, show a 45% decline in U.S. production over the past 10 years (Fig. 1); the average wholesale value has risen during the same time period. The production figures are somewhat slanted and do not present an accurate picture of North American hydrangea production. Only 28 states were polled in the census, and Canadian production, which has increased steadily over the past 10 years (90), is not included in the figures. Although a decline in U.S.

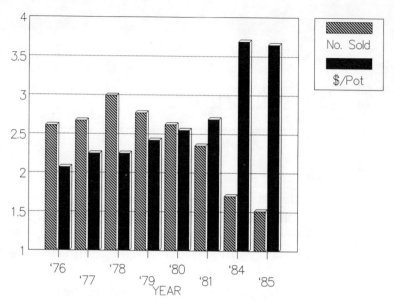

Figure 1. Number (millions) and average sale price ($/pot) of wholesale potted hydrangeas produced in the U.S. from 1976 to 1985. Based on U.S. Census of Floriculture Crops; 28 states participating (77–84).

production has occurred, it is not to the extent portrayed by the census figures—probably only about 20%. The decline is likely related to more efficient cropping technology for competing crops. Little research has been conducted on streamlining the production of hydrangeas compared to crops such as chrysanthemums and Easter lilies. The average production time for potted chrysanthemums (from unrooted cuttings) is about 12 weeks; Easter lilies are in the greenhouse for 16–18 weeks. Conventional hydrangea production may take more than 52 weeks from time of sticking cuttings to sale for Mother's Day. Hydrangeas are not an easy plant to grow, and their decline is due more to production difficulties than reduced consumer popularity.

The "conventional florists' hydrangea" is 12–20 in. tall, bears 1–4 inflorescences, and is planted in a 6–8 in.

diam. pot. Over the past 5–10 years, two alternative hydrangea forms have increased in popularity: 1) multi-stem plants with 5 or more cymes and 2) mini-pot hydrangeas (59). The multi-stemmed plants have more but smaller flower heads compared with the conventional hydrangea. Mini-pot hydrangeas are 7–10 in. tall, have one small (4–6 in. diam.) flower head, and are produced in 4–5 in. diam. pots. Both products appeal to a different market than conventional hydrangeas and have the potential to increase hydrangea popularity in the future.

This manual is designed to reduce the difficulties encountered by hydrangea growers and expand future hydrangea production. I encourage all comments and welcome any suggestions. All input will be incorporated in later editions. This way, we can continue to improve technology for hydrangeas, reduce growing time, and bring hydrangea production into the "fast lane."

Cultivars

There are over 500 hydrangea cultivars available today, but only a few are commercially important in North America (Table 1). Successful cultivars must consistently perform well for the grower as well as maintain popularity with the consumer. The popularity of a cultivar differs among climatological regions. For example, 'Todi' is popular with Canadian growers due to durability in a colder climate and its earliness in flowering. However, most southern U.S. growers prefer 'Rose Supreme' because of heat tolerance during forcing. Cultivars differ in sepal colors and market preference because color often influences cultivar popularity from region to region. The targeted sales date also may dictate cultivar. Slower, late flowering cultivars are better suited for Mother's Day, while producers desiring sales for an early Easter would grow an early flowering cultivar. The final consideration for cultivar selection is plant form desired. All top 15 cultivars in Table 1 are suitable for production as conventional plants with 1–4 cymes each. However, I recommend 'Kasteln', 'Mathilda Gutges', 'Bottstein', 'Sister Therese', 'Blau Donau', and 'Dr. Bernard Steiniger' as the best cultivars to form multi-stemmed plants because of their smaller cymes and self-branching habit. For mini-pot production, 'Merritt's Supreme', 'Blau Donau', and 'Dr. Bernard Steiniger' have been successful. These cultivars force rapidly and are attractive as small, single-stemmed plants.

Table 1.	Hydrangea cultivars by popularity in North America. Adapted from (24, 28, 34, 90).				

Popularity in North America	Cultivar	Sepal color description[a]		Relative days to flower at 60°F night temperature[b]	Inflorescence size/ plant height	Comments
		Grown as a pink	Grown as a blue			
1	Merritt's Supreme	Deep pink	Medium blue	88	Large/Short	Heat tolerant
2	Todi (Toddy)	Dark pink	Not recommended	88	Large/Medium	Not heat tolerant
3	Rose Supreme	Rose pink	Light blue	103	Very Large/Tall	Heat tolerant
4	Kasteln	Deep pink	Medium blue	95	Small/Medium	
5	Mathilda Gutges	Pink	Light blue	95	Medium/Medium tall	
6	Bottstein	Light red	Not recommended	92	Medium/Medium	
7	Kuhnert	Not recommended	Medium blue	95	Large/Tall	
8	Sister Therese	White	White	95	Medium/Tall	Best white color
9	Red Star	Brilliant red	Medium blue	95	Large/Tall	
10	Schenkenburg	Dark red	Not recommended	95	Medium/Medium	Best red color
11	Wildenstein	Pink	Not recommended	95	Medium/Medium	
12	Brestenburg	Not recommended	Dark blue	95	Medium/Tall	Best blue color
13	Strafford	Dark pink	Not recommended	99	Large/Medium	Not heat tolerant
14	Enziandom	Dark pink	Deep blue	92	Medium/Medium	
15	Merveille	Light red	Not recommended	95	Large/Medium	

Regula	White	White	88	Medium/Medium tall	
Blau Donau	Pink	Clear blue	95	Medium/Medium	Heat tolerant
Rosa Rita	Light pink	Light blue	88	Medium/Tall	Not heat tolerant
Dr. Steiniger	Rose red	Not recommended	83	Medium/Medium	Earliest bloomer
Charme	Bright red	Not recommended	88	Large/Tall	

[a]The sepal color of hydrangeas (except for whites) can be modified through soil pH and nutrition manipulations. Color descriptions are given for plants grown for both pink or blue sepals.

[b]Based on using a 60°F night temperature up to 18 days prior to sale, then lowering night temperature to 54°F for color intensification. The amount of time required during forcing depends on temperature and light conditions.

Propagation

Plant Material

Although it is possible to maintain stock plants, hydrangeas are produced almost exclusively from unrooted softwood cuttings obtained from the West Coast. The cuttings are shipped as 5–7 in. long (3–4 nodes each) terminals from late April through early July. The original cuttings are re-cut by the grower into a 2–4 in. long terminal cutting and a single two-eye (butterfly) cutting or two single-eye (leaf-bud) cuttings. Single-eye cuttings are made by slicing through the shoot between the points of attachment for the leaf petioles. Each type of cutting has advantages and disadvantages, outlined in Table 2. Most production emphasizes terminal (tip) cuttings, mainly due to the reduction in propagation and establishment time as compared to butterfly or single-eye cuttings.

Rooting Media

Many propagation media are acceptable for hydrangeas. Growers have used sand, sand / peat moss, vermiculite, vermiculite / perlite, soil / peat / perlite, bark / peat / perlite, or perlite successfully. The important parameters are adequate water-holding capacity, good drainage, a high degree of porosity, and absence of any rot-causing pathogens. Rooting media should be steam pasteurized

Table 2.	Comparison of cutting types.
Cutting type	*Comparative advantages and disadvantages*
Terminal tips	—root and establish more rapidly than butterfly or single-eye cuttings —require more propagation area than butterfly or single-eye cuttings
Butterfly	—give the grower two shoots per cutting very early in production
Single-eye	—make the most efficient use of propagation material —are more susceptible to disease during propagation than butterfly or tip cuttings (90) —require about 4 weeks longer than terminals to develop a shoot of comparable size

prior to use and void of pesticide residues. Hydrangeas should not be rooted in media treated with Terraclor® (quintozene) as growers have reported significant reductions in hydrangea rooting in the presence of this fungicide. Banrot® (etridiazole plus thiophanate-methyl) also may delay rooting (90). Benlate® and Tersan 1991® (benomyl) do not hinder hydrangea rooting and may be used if needed for control of root-rot organisms (90).

Cuttings may be direct rooted in 2¼–4 in. pots (28,76); "direct stick" technique requires use of more propagation mist area, and the root medium should be able to support plant growth until transplanting later in the summer. Two systems (Oasis® Rootcubes® and Jiffy-7®'s) which work well for other floriculture crops have been reported to result in low survival rates of cuttings (90), probably due to poor aeration and drainage in the media.

Rooting Hormones

Rooting can be increased with the use of 10,000 ppm IBA in talc (19). This formulation is approximated by mixing equal parts of "Hormo.-Root 2" (2% IBA) with talc. Scented baby powder should not be used because the fragrances used could reduce rooting. Complete mixing of dry powders can be assured by making a slurry with 91% isopropyl alcohol (or any high proof, clear, distilled alcohol) and letting the material dry overnight in a shallow tray. The cake is easily broken into fine powder for use. Over-application of cuttings with an IBA-talc formulation results in damaged tissue, easily attacked by pathogens. Therefore, be sure cuttings retain only a light layer of powder prior to sticking. A 'quick-dip' into an aqueous solution containing 250 ppm NAA and 500 ppm IBA is equally effective as the IBA-talc treatment (90). This solution can be made by mixing 1 part "Dip'n Grow" with 19 parts water. If a liquid treatment is used, change the solution frequently to prevent possible spreading of pathogens from contaminated to clean cuttings. A 2500 ppm aqueous solution of B-Nine® (daminozide) also has been reported to aid rooting (29), but the possibility of the B-Nine® delaying bud break after propagation has not been investigated fully.

Propagation Environment

IBA-treated cuttings should be placed in a propagation medium maintained at 72°F. Intermittent mist should be used to prevent wilting. High pH water and/or hard water can lead to mineral deposits on the leaves, and retard the growth of the cutting (76) which a pH of 6.0 can prevent. High pH can be reduced by injecting phosphoric or sulfuric acid into the system. Shade cuttings to 2500–3000 ft-c in high light areas to reduce leaf temperatures during propagation. Do not to cover the buds of cuttings with the leaves of adjacent cuttings, a sure way to stress cuttings and kill the apex. Hydrangea leaves are large, so many propagators remove half of each leaf, allowing more cuttings per square foot and reducing the possibility of covering buds. Removal of too much leaf tissue during propagation, however, slows post-rooting establishment due to reduced food (carbohydrate) reserves in the plant.

Depending on cultivar and type of cutting, 3–5 weeks are required for adequate rooting. Cuttings must be potted as soon as an adequate root system is formed, as maintaining cuttings under mist longer retards establishment and development of new leaves. If plants cannot be potted immediately, turn off mist and water periodically with a dilute fertilizer solution (50 ppm N from a 2N-1P-1K source). If cuttings do not start to grow by the second week after transplanting, spray plants with a dilute (5 ppm) GA_3 solution (Bailey, unpublished data). These first weeks after potting are crucial in establishing strong shoots: delays in shoot development severely reduce plant quality.

Maintaining Stock Plants

Although most growers buy cuttings, some maintain their own stock and have vegetative cuttings available on a year-round basis. The following 3-point plan should be followed to maintain vegetative stock plants: 1) keep plants under a 24-hr. photoperiod (dusk-to-dawn lighting) in a 70°F minimum greenhouse (12); 2) remove stem tips every 60 days; and 3) spray plants with 25 ppm GA_3 2–5 days after pinching (59). Use large pots (3–5 gallons) to permit rapid growth of stock plants. Hydrangeas are shrubs, so stock plants eventually produce woody growth unsuitable for cutting production. Stock should· be replenished every 18–30 months to assure maximum cutting production and reduce disease.

Micropropagation

Hydrangeas may also be propagated by tissue culture. To date, only shoot tip and apex cultures have been utilized (20,27,64). Cooperative research between Washington State and Oregon State universities has resulted in virus-free hydrangeas from shoot apex culture (4). Growers interested in micropropagation of hydrangeas should acquire the above references for more information.

Plant Establishment and Vegetative Growth

In conventional hydrangea production, the vegetative growth phase occurs during the summer months either in greenhouses or outdoors under lath or shade cloth. The goals for this phase of production are to 1) establish the rooted cuttings; 2) determine which sepal color (pink or blue) is to be produced upon forcing; 3) stimulate production of multiple shoots, each strong enough to produce an inflorescence; and 4) maintain short internodes and prevent excessive shoot height.

Valuable insight into the environmental needs of hydrangeas may be gained by examining their native habitat. Florists' hydrangeas are native to the central Pacific coast of the Japanese island of Honshu (25,45,88). This area has more than 5 months of frost-free temperatures. The mean low and high temperatures are 31°F and 47°F in January and 72°F and 85°F in August, respectively. Annual rainfall is between 70 and 90 in. while mean relative humidities during January and August are 65% and 85%, respectively (6,46). These conditions provide growers very important production guidelines—hydrangeas are accustomed to moderate temperatures, moderate-to-high humidity, and *an abundant supply of water.*

Figure 2. Well-rooted hydrangea cutting, ready for potting.

Table 3.   **Growth media for hydrangeas. The component volumes given per yd^3 and m^3 allow for approximately 17% shrinkage upon mixing.**

Ingredient	Amount/yd^3 of mix	Amount/m^3 of mix
Pink or White Flowers		
Top soil (Sandy loam)	6.5 ft^3 (5 bushels)	240 liters
Sphagnum peat moss	13 ft^3 (10.5 bushels)	480 liters
Perlite	13 ft^3 (10.5 bushels)	480 liters
Treble superphosphate	1.5 lbs	890 g

Ingredient	Amount/yd³ of mix	Amount/m³ of mix
Potassium nitrate or Calcium nitrate	1 lb	593 g
Ground dolomitic limestone OR[a]	0–8 lbs (raise pH to 6.2)	0–4.75 kg
Aluminum sulfate	0–5 lbs (lower pH to 6.2)	0–4.75 kg
Calcium sulfate (Gypsum)[b]	0–5 lbs	0–2.97 kg
Magnesium sulfate (Epsom salts)[b]	0–24 ozs	0–890 g
Fritted trace elements 555	2 oz	74 g

Blue Flowers

Ingredient	Amount/yd³ of mix	Amount/m³ of mix
Top soil (Sandy loam)	6.5 ft³ (5 bushels)	240 liters
Sphagnum peat moss	13 ft³ (10.5 bushels)	480 liters
Perlite	13 ft³ (10.5 bushels)	480 liters
Potassium nitrate or Calcium nitrate	1 lb	593 g
Ground dolomitic limestone OR[a]	0–5 lbs (raise pH to 5.5)	0–4.75 kg
Aluminum sulfate	0–5 lbs (lower pH to 5.5)	0–4.75 kg
Calcium sulfate (Gypsum)[b]	0–5 lbs	0–4.75 kg
Magnesium sulfate (Epsom salts)[b]	0–24 ozs	0–890 g
Fritted trace elements 555	2 oz	74 g

[a] Test a sample of mix pH then adjust if needed to reach the recommended pH. Add dolomitic limestone to raise the pH to the desired value *or* add aluminum sulfate to lower the pH to the desired value.
[b] The amount added should be inversely proportional to the amount of dolomitic limestone used. For example, if no dolomitic limestone is needed for pH control, incorporate 5 lbs/yd³ of calcium sulfate and 24 oz/yd³ of magnesium sulfate to assure adequate Ca and Mg.

Growth Media and Potting

Cuttings should be potted as soon as an adequate root system has been produced, as rapid establishment of cuttings is the first objective. Hydrangeas are ready for transplant when they have a root system similar in size to the plant shown in Figure 2. The time required to reach this stage varies with type of cutting, cultivar, and time of year, but in general, such a root system is adequate for cutting survival and rapid establishment. Newly transplanted cuttings appreciate afternoon syringing and moderate shading (less than 4000 ft-c) to reduce transplant shock. Plants should recover after 3–7 days (i.e. leaves turgid) after which shading is removed.

The choice of growth medium is important for hydrangea production. Plants may remain in the same container for 22 weeks or more, so the mix is critical for proper growth. Hydrangeas require abundant water, so the growth medium must be selected for good moisture-holding capacity. A peat moss-based medium affords good moisture retention; field soil, perlite, vermiculite, pine bark, hardwood bark compost, or other available materials can be used in combination with peat moss. We have had success (in Indiana and Arizona) with a 1 soil: 2 sphagnum peat: 2 perlite mix by volume (Table 3). Regardless of the recipe used, the medium's physical and chemical properties should fall within the ranges outlined in Appendix B.

Plants are normally maintained in 4 in. pots during the vegetative growth phase. This system takes up less space, but close monitoring of water is required to prevent wilting.

Control of Sepal Color

The sepals of non-white cultivars can be pink, blue, or any number of shades of mauve. The crucial time for sepal color regulation begins during the vegetative growth phase. Growers achieve better "blues" if the bluing program is begun at time of transplant in the summer (9,22,29). Begin the "pinking" or "bluing" program as early in production as possible to assure clear sepal coloration upon forcing. Table 1 lists cultivar color descriptions as well as suggested best blues and pinks.

The environmental factor controlling the color of the sepals is the availability of aluminum in the growth medium (3). Pink and blue inflorescences contain the same anthocyanin pigment, namely delphinidin 3-monoglucoside (7,55). When aluminum is present in the sepals, it binds with the pigment and a co-pigment, 3-caffeoylquinic acid. The sepal color changes from pink to blue when aluminum is available to bind with the pigment and co-pigment (70). Conditions favoring an increase in aluminum availability lead to blue flowers; low availability of aluminum results in pink flowers. Poorly controlled conditions, or those left to chance often result in undesirable lavender blends. Aluminum availability is related to the pH of the soil solution and to plant nutrient levels within the soil solution; most importantly, phosphorus as phosphorus drastically reduces the availability of aluminum.

For pink sepals, the soil solution should have the following characteristics:
1) Low levels of aluminum (<15 ppm) (1). If aluminum is unavailable, the plants cannot absorb enough to produce the blue pigment complex.
2) A high level of nitrogen (30–50 ppm NO_3-N [Spurway extraction]) (22). High levels of nitrogen have been correlated with clear pink coloration (28,38). High

nitrogen levels in the soil antagonize the uptake of aluminum, thus helping to prevent blue sepals (10). High ammonium (NH_4-N) relative to nitrate (NO_3-N) nitrogen also helps produce pink sepals (8). Aluminum uptake apparently is reduced to a greater degree by NH_4-N than by NO_3-N. However, even high ammonium levels are not sufficient to produce pink sepals if high levels of aluminum are present.

3) A high level of phosphorus (6–12 ppm [Spurway extraction]) (9,61). Phosphorus antagonizes aluminum uptake, and high concentrations prevent blue coloration of sepals.

4) A low level of potassium (10–20 ppm [Spurway extraction]) (10,66). Potassium is involved in the change from pink to blue. A low level of potassium helps assure clear pink sepals.

5) A pH of 6.0–6.5 (22,76). In this pH range, aluminum is marginally available, further assuring that sepals will be pink.

In general, the following soil solution leads to blue sepals:

1) An ample supply of aluminum (>100 ppm) (2,3). To assure adequate aluminum, drench blue plants during September with aluminum sulfate (15 lbs/100 gallons). Make 2 applications, 14 days apart. See *Forcing Blue Plants* in section "Greenhouse Forcing" for details on applying aluminum sulfate drenches.

2) A medium level of nitrogen (20–30 ppm NO_3-N [Spurway extraction]) (22). Although nitrogen appears to antagonize aluminum uptake, hydrangeas require a moderate nitrogen level for best growth and attractive foliage.

3) A low level of phosphorus (1–5 ppm [Spurway extraction]) (61,66). Aluminum uptake by the plant is more efficient with reduced phosphorus levels in

the soil. It is essential for clear blue sepals to limit soil phosphorus.

4) A high level of potassium (25–50 ppm [Spurway extraction]) (38,66). High potassium levels are associated with clear blue coloration of sepals.

5) A pH of 5.0–5.5 (1,52). In this pH range, aluminum is more available than at a higher pH, encouraging additional aluminum uptake.

Soil surveys of the Honshu peninsula report most areas have an acidic reaction (40,69), and florists' hydrangeas develop blue sepals in their native habitat. Ironically, the majority of hydrangeas sold in North America have pink inflorescences.

Nutrition and Fertilization

Hydrangea nutrition is obviously critical for both optimum plant growth and controlled sepal coloration. Programs for pink sepals differ from those recommended for blue sepals (Table 4). These recommendations are designed to produce the sepal coloration conditions presented above. White cultivars can be produced under either program but are usually grown using nutrition guidelines for pink flowers. The fertilization recommendations given in Table 4 are not the only feasible methods available and are given as suggestions only. If an alternate fertilization program is chosen, however, it must fulfill the requirements outlined for the sepal color intended.

Hydrangeas are susceptible to iron deficiency chlorosis, especially at pH above 6.0. Drenches of iron chelate are effective in controlling the problem. Use 3–5 oz of 10% chelated iron per 100 gallons and provide enough drench to each pot to achieve run through. Multiple applications every 4–8 weeks may be necessary to prevent recurrence. Applications should be made as soon as pos-

Table 4.	Liquid fertilization programs for pre-forcing production.

Pink or White Flowers

Continual fertilization. Supply 100–175 ppm N at each irrigation using:
 a. 5.3–9.3 oz 25-10-10 per 100 gallons *or*
 b. 1–2 oz 11-53-00 (monoammonium phosphate) plus 1–2 oz 13-00-44 (potassium nitrate) plus 3.5–6 oz 32-00-00 (Uran 32) per 100 gallons

Maintain irrigation solution pH at 6.3 using phosphoric, sulfuric, or citric acid. Phosphoric acid is preferred to maintain desired levels of P for pink flowers.

Intermittent fertilization. Supply 340–450 ppm N every 7–10 days using:
 a. 18.2–24 oz 25-10-10 per 100 gallons *or*
 b. 3.5–4.5 oz 11-53-00 (monoammonium phosphate) plus 4–5.5 oz 13-00-44 (potassium nitrate) plus 11–15 oz 32-00-00 (Uran 32) per 100 gallons

Maintain fertilizer solution and clear water pH at 6.3 using phosphoric, sulfuric, or citric acid. Phosphoric acid is preferred to maintain desired levels of P for pink flowers.

Blue Flowers

Continual fertilization. Supply 100–175 ppm N at each irrigation using:
 a. 6.7–11.7 oz 20-5-30 per 100 gallons *or*
 b. 0.5–1 oz 11-53-00 (monoammonium phosphate) plus 4.5–7.5 oz 13-00-44 (potassium nitrate) plus 2.5–4 oz 32-00-00 (Uran 32) per 100 gallons

Maintain irrigation solution pH at 5.3 using sulfuric or citric acid. Do *not* use phosphoric acid; the added P is counterproductive.

Intermittent fertilization. Supply 340–450 ppm N every 7–10 days using:
 a. 22.7–30 oz 20-5-30 per 100 gallons *or*
 b. 2–3 oz 11-53-00 (monoammonium phosphate) plus 15–20.5 oz 13-00-44 (potassium nitrate) plus 7.5–9.5 oz 32-00-00 (Uran 32) per 100 gallons

Maintain fertilizer solution and clear water pH at 5.3 using sulfuric or citric acid. Do *not* use phosphoric acid; the added P is counter-productive.

22

sible if interveinal chlorosis appears on newly expanding leaves. Although hydrangeas are susceptible to other nutrient deficiencies (Tables 5 and 6), an adequate fertilization program should prevent deficiencies.

The pH of the growth medium is important for proper sepal coloration and nutrient uptake. If little or no dolomitic limestone is added to the media, calcium sulfate and magnesium sulfate should be incorporated to supply adequate Ca and Mg (Table 3). Also, calcium nitrate should be incorporated in the soil recipe instead of potassium nitrate. If calcium levels in the soil solution drop too low (less than 100 ppm [Spurway extraction]), calcium nitrate should be utilized occasionally as a nitrogen source. Do not mix calcium nitrate with phosphoric acid or phosphate-containing fertilizers, as calcium phosphate precipitates out of solution and thus is not available.

Table 5.   **Descriptions of nutrient deficiency symptoms on hydrangea.** Adapted from (13, 15).

Deficiency	Visible symptoms
Nitrogen (N)	—Older, lower leaves chlorotic with necrotic tips. —Red margins apparent on the young expanding leaves. —Few of the buds develop and shoot elongation is reduced. —Stems have distinct purple tint as do the bud scales.
Phosphorus (P)	—Older leaves have purple coloration along the margins. —Older leaves slightly chlorotic. —Internodes extremely reduced. —Young foliage smaller than normal and very dark blue-green.
Potassium (K)	—Young leaves are very dark green. —New leaves are narrower than normal and appear more lustrous. —Shoots are telescoped, giving a rosette appearance.

Deficiency	Visible symptoms
Calcium (Ca)	—Young leaves appear thin and very translucent with black, necrotic tips. —Apices almost completely necrotic and blackened.
Magnesium (Mg)	—Root system is reduced in size, yet appears healthy. —Oldest leaves interveinally chlorotic with red margins.
Sulfur (S)	—Young leaves generally chlorotic, more so along the margins. —Internodes are reduced in length.
Boron (B)	—Root tips are tan instead of white; side roots are stubby and brown. —Young leaves chlorotic, especially at the base of secondary veins. —Chlorotic regions acquire a tan hue with time.
Iron (Fe)	—Severe interveinal chlorosis mainly on young leaves. —Necrotic areas prevalent along the young leaf margins.
Zinc (Zn)	—Young leaves distorted and slightly chlorotic. —Newly expanding leaves puckered and curl inward towards the apices.

Table 6. **Foliar nutrient levels when deficiency symptoms appear in 'Rose Supreme' hydrangeas.** Adapted from (13, 15).

Macronutrients	(% dry weight)	Micronutrients	(ppm)
N	≤ 1.50	B	≤ 25
P	≤ 0.20	Fe	≤ 60
K	≤ 1.00	Mn	< 8
Ca	≤ 0.60	Cu	< 4
Mg	≤ 0.08	Zn	≤ 20
		Mo	< 1

Watering

Hydrangeas have a high demand for water, and no other plant is as unforgiving of wilt as hydrangea. Their large leaf area results in large water losses in relatively short periods of time. Leaf margins burn and die if plants are allowed to wilt. Enough water must be applied to each pot to result in 10% run-through at each watering, e.g. if 10 fluid ounces is required to saturate a pot, 11 ounces should be applied.

Just as wilt should be avoided, over-watering must also be avoided. Over-watering is not the result of applying too much water at each irrigation, but rather results from watering too frequently. Maintain an ample moisture supply at all times and avoid water stress of any kind.

Water quality should be assessed and adjusted prior to use. Total soluble salts as measured with a solubridge should be below 1.75 dS $\cdot$ m^{-1} (mmho $\cdot$ cm^{-1}). The pH of the irrigation solution should be adjusted as described previously. Water quality should be checked periodically during production to assure proper pH regulation. Many municipalities change their source of water throughout the year and, therefore, the water quality (mainly pH) can also change. The pH control program should be adjusted accordingly.

Light Intensity and Temperature

Hydrangeas require adequate light intensity during vegetative growth to produce strong plants with ample branching. Optimum light levels vary with the plant's stage of growth. Propagation should be conducted under 2500–3000 ft-c. Newly transplanted cuttings should receive 3000–4000 ft-c for about a week after transplanting to reduce transplant shock. After the plants have been established, light should be increased to between 5000 and 7500 ft-c.

Although best plant growth occurs under the higher light recommended, shading to 5000 ft-c may be necessary to control plant temperature. If plants are subjected to high temperatures (greater than 86°F) for extended periods of time, they acquire symptoms of "hydrangea distortion" (see section "Pests, Diseases, and Other Problems") (13,16). Cooling the plants through heavier shading, ventilation, fan and pad cooling, or mist cooling helps reduce the risk of hydrangea distortion. Fans and pad pumps should be set to turn on when the interior temperature reaches 80–82°F.

Minimum temperatures are equally important for vegetative growth. Hydrangeas initiate inflorescences more readily and vegetative growth is reduced as night temperatures reach 70–65°F (39,47). If night temperatures in late May to late August fall below 70°F for more than 5–10 consecutive nights, summer production should not be attempted outdoors. Plants should be grown in a greenhouse, where night temperatures can be maintained at or above 70°F to prevent premature inflorescence initiation. Night temperatures higher than 72°F but lower than 75°F are best in assuring vegetative plants, but are rarely economical in a greenhouse.

Photoperiod

Growth responses of hydrangeas are closely related to both photoperiod and temperature. In general, long days promote vegetative growth whereas short days promote floral initiation. However, temperature interacts with photoperiod. For example, with warm night temperatures (above 70°F), floral initiation can be delayed by increasing the daylength (39,47). Greater daylength results in greater delay, and a 24 hr. photoperiod (dusk-to-

26

dawn lighting) is most effective in maintaining vegetative growth (17). Table 7 lists the more important photoperiod/temperature combinations and the plant response to each combination.

Table 7. **Summary of hydrangea responses to night temperatures and photoperiod.**
Adapted from (12, 17, 39, 47).

Night temperature (T)	Photoperiod (hrs per day)	Plant response
T ≥ 80°F (T ≥ 27°C)	8–24	Vegetative; could damage inflorescence primordia initiated previously and possibly lead to hydrangea distortion
80° > T ≥ 70°F (27° > T ≥ 21°C)	24	Strongly vegetative; floral initiation very unlikely
80° > T ≥ 70°F (27° > T ≥ 21°C)	8	Floral initiation stimulated
70° > T ≥ 65°F (21° > T ≥ 18°C)	24	Delays (does not prevent) floral initiation
70° > T ≥ 65°F (21° > T ≥ 18°C)	8	Floral initiation stimulated
65° > T ≥ 52°F (18° > T ≥ 11°C)	8–24	Floral initiation occurs readily
52°F > T (11°C > T)	8–24	Little floral initiation occurs; buds go dormant

Pinching

One of the objectives during the summer vegetative growth phase is to produce multiple shoots per plant, each strong enough to develop a full inflorescence. Hydrangeas can be grown with single, double, or multiple inflorescences per plant. The number of inflorescences depends on the cutting material utilized and the pinching program implemented (Table 8). For multiple inflorescence plants, the first pinch, leaving 2 nodes per shoot, should be made after the newly potted cutting has been established, about 2–3 weeks after potting. If a second pinch is to be made, allow the shoots arising from the first pinch to develop so that each will have 2 nodes remaining after the second pinch. The final pinch should be made no later than July 5 to assure ample shoot growth prior to floral initiation. Pinching later than early July often results in reduced inflorescence size and a higher percentage of blind (non-flowering) shoots upon forcing (59).

After pinching has been completed and prior to floral initiation in late August, growers should selectively prune weak, thin-stemmed shoots to reduce potential for blind shoots (28). Only strong, thick shoots must be allowed to develop fully.

Table 8. **Summary culture schedules based on cutting type for dormant plant production (cold storage starting in early November).** Adapted from (59, 86).

Desired no. of inflorescences per plant	Final pot size at forcing[a]	Propagation date	Date of first pinch	Date of second pinch
Tip cuttings				
1	5½–6 in.	6/7–6/14	—	—
3–4	7–8 in.	5/7–5/14	6/21–6/28	—
>4	8 in.	4/17–4/23	5/29–6/4	6/26–7/2
Butterfly cuttings				
2	6–7 in.	5/7–5/14	—	—
3–4	7–8 in.	4/24–4/30	6/26–7/2	—
Single-eye cuttings				
1	5½–6 in.	5/7–5/15	—	—
2–4	6–8 in.	4/10–4/16	6/26–7/2	—

[a]'Rose Supreme' plants require the larger size pots. The smaller size given for each plant type is appropriate for standard cultivars.

Chemical Growth Retardants

The final objective of the summer vegetative growth phase is height control. Hydrangeas are shrubs and naturally grow to heights of 3–7 feet in their native habitat. Height control is important during vegetative growth and forcing. Not only is height control important aesthetically, but excessive internode elongation prior to floral initiation results in a higher percentage of blind shoots as well (29).

B-Nine® (daminozide) has been the most effective and widely used material for height control. Plants should be sprayed after they have received their final pinch and when the newly developing shoots are 1–1½ in. long (28,76). Spray concentrations of B-Nine® used are 5000–7500 ppm. A second and third application made at 2–3 week intervals may be needed for effective height control, especially for tall-growing cultivars like 'Rose Supreme'.

Other growth retardants are also effective for height control: A-Rest® (ancymidol) sprays (50–100 ppm) (21,74), A-Rest® drenches (2–4 mg a.i./pot) (74), Cycocel® (chlormequat) drenches (0.3–1.6 g a.i./pot) (5), Bonzi® (paclobutrazol) sprays (50–100 ppm) (21,56), and Sumagic® (XE-1019, uniconazole) sprays (5–20 ppm) (14). However, more work is needed in "fine-tuning" the concentrations of Bonzi® and Sumagic®. Cycocel® sprays are not effective on hydrangeas (21,32).

Table 9 provides growth retardant spray recipes for use on hydrangeas. The growth retardant spray concentrations given are based on an application volume of 1 gallon per 200 ft^2 of growing area (71). If a different spray volume per unit growing area is to be used, adjust the growth retardant concentration accordingly. Apply the sprays evenly over the entire growing area to assure uniform coverage and plant response. Table 10 gives drench recipes for effective growth retardants. The media should be moist prior to drenching to prevent root damage. Drench volume should be 1–1½ fl oz per inch of pot size (diameter).

Height control should be continued through late August, prior to floral initiation. Although B-Nine® applications during the summer do not affect the resulting inflorescence size (57), applications made late in the fall will have carry-over effect in terms of shoot length upon forcing. This is not undesirable, but should be taken into account for the height control program during the forcing period.

Table 9. Growth retardant spray solutions for hydrangeas.

Chemical	Spray solution (ppm)	Amount/gallon of final solution	Amount/liter of final solution
B-NINE SP®	2500	0.39 oz/gal	2.94 g/liter
(850 g a.i./kg)	5000	0.79 oz/gal	5.88 g/liter
	7500	1.18 oz/gal	8.82 g/liter
A-REST®	50	22.24 fl oz/gal	189.4 ml/liter
(0.264 g a.i./liter)	100	48.48 fl oz/gal	378.8 ml/liter
BONZI®	50	1.6 fl oz/gal	12.5 ml/liter
(4 g a.i./liter)	100	3.2 fl oz/gal	25.0 ml/liter
SUMAGIC®	5	1.28 fl oz/gal	10 ml/liter
(0.5 g a.i./liter)	10	2.56 fl oz/gal	20 ml/liter
	20	5.12 fl oz/gal	40 ml/liter

Table 10. Growth retardant drench solutions for hydrangeas.

Chemical	Dose (mg/pot)	Drench vol. per pot	ppm	Amount/gallon of final solution	Amount/liter of final solution
A-REST®	2	8 fl oz	8.5	4.10 fl oz/gal	32 ml/liter
(0.264 g	3	8 fl oz	12.7	6.15 fl oz/gal	48 ml/liter
a.i./liter	4	8 fl oz	16.9	8.20 fl oz/gal	64 ml/liter
CYCOCEL®	300	6 fl oz	1691	1.83 fl oz/gal	14.3 ml/liter
(118 g	900	6 fl oz	5072	5.50 fl oz/gal	43.0 ml/liter
a.i./liter	1600	6 fl oz	9018	9.78 fl oz/gal	76.4 ml/liter

Floral Initiation

Initiation of inflorescences follows summer vegetative growth. Floral initiation in hydrangeas is complex and affected by light intensity, water status, nitrogen fertility, plant size, night temperature, and photoperiod.

Photoperiod and Temperature Responses

Hydrangeas initiate and develop inflorescences during autumn in their native habitat (33,43). Both temperature and photoperiod are involved in floral initiation (Table 7). Although daytime temperature and daily temperature averages (average of 24 hour minima and maxima) are important, nighttime temperature appears most closely associated with floral initiation (51,62). A night temperature of 52–65°F is most desirable for floral initiation in hydrangeas (26,47,62). Short photoperiods (8–12 hours per day) are more conducive to floral initiation than longer photoperiods (48,49,68). As shown in Table 7, 24 hr. photoperiods delay floral initiation at night temperatures greater than 65°F; and at night temperatures above 70°F, 24 hr. photoperiods prevent floral initiation (17,47). Depending on the cultivar, 6–10 weeks of cool, short days are needed to provide sufficient development of floral primordia prior to cold storage (see section "Defoliation and Dormancy" for discussion of cold storage) (28,34).

Ideally, night temperatures should be maintained at 52–65°F to assure rapid floral initiation. If night temperatures are higher, stimulate inflorescence formation with 8 hour photoperiods by covering plants with blackcloth from 4:00 PM–8:00 AM. Do not allow the night temperature under the blackcloth to reach 80°F or floral initiation will be delayed, distorted, or aborted.

Cultural Practices During Floral Initiation

Hydrangea floral initiation is affected by factors other than photoperiod and night temperature.

Light intensity is important in the production of strong shoots capable of producing large inflorescences. Light levels should be maintained above 2000 ft-c during the floral initiation phase to encourage strong shoots. Below 2000 ft-c, floral initiation slows and blind shoots increase (39,54). Plants should not be crowded; space them far enough apart (8 × 8 in.) to prevent shading of adjacent plants. High light intensity has been reported to prevent floral initiation (39); however, this appears to be a result of increased plant temperature and is not a direct effect of high light. If daytime temperatures are greater than 86°F during autumn, shading to 5000 ft-c is beneficial in reducing plant temperature, and floral initiation occurs more readily.

Reduced nitrogen (N) fertilization stimulates hydrangea flower bud formation (23). Nitrogen should be decreased concomitant with the onset of autumn. Reduce the N level by half during September and cease fertilizer application during October (76).

Stages of Inflorescence Development

Growers should follow the development of the cyme inflorescence. Dissect 2 or 3 apices every 2 weeks starting in early September in order to follow the initiation process. Although it is possible to view apices with a hand-held 10× lens, a stereoscopic microscope is a worthwhile investment for this purpose.

Vegetative hydrangea apices are narrow and are covered by the upper leaf pair primordia (Figure 3, Stage 1). As the floral initiation process begins, the apex broadens (Figure 3, Stages 2 & 3), and internode expansion slows down. Finally, 5 primary inflorescence primordia become visible (Figure 3, Stage 4). Each of the primary axes develop 3 secondary axes (Figure 3, Stage 5). In most instances, each of the secondary axes form 3 tertiary axes. The individual fertile flowers are borne on the secondary and tertiary axes (Figure 3, Stage 6). The individual fertile flower primordia should be visible prior to placement into cold storage. Sepal and perhaps petal primordia on each flower primordia are visible with a 10× hand lens. After apices have reached this stage of development, plants can be placed into cold storage (see section "Defoliation and Dormancy"). The timing of floral initiation is dependent on location and on the cultivar being grown. In general, floral initiation becomes visibly evident (apices at stage 3) in mid-September at 40° N latitude. By early- to mid-November, plants are ready for placement into cold storage.

Further differentiation of the fertile flowers and the development of the sterile flowers does not take place until the plants are placed into a warm greenhouse, during the forcing phase of production (23;89). The sterile flowers that compose the majority of the inflorescence are not present until the forcing stage. The fertile flower primordia that are easily observable prior to cold storage will eventually be buried beneath the larger, sterile flowers upon full expansion of the inflorescence.

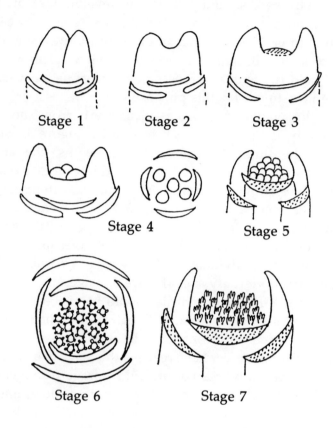

Figure 3. Stages of hydrangea inflorescence development adapted from (39). Stages: 1) Apex is completely covered by the upper leaf pair primordia; meristem is vegetative. 2) Apex broadens, upper leaf pair primordia separates. 3) Apex swells and becomes dome-shaped. 4) Five primary inflorescence axis primordia are visible. 5) Both primary and secondary inflorescence axis primordia are visible. 6) Sepal and petal primordia are visible for individual fertile flowers. 7) Sepal, petal, stamen, and pistil primordia are visible for individual fertile flowers.

Chemical Stimulation of Floral Initiation

Maintaining a 52–65°F night temperature is an effective method of stimulating hydrangea flower initiation, but 52–65°F is not always attainable so chemical growth retardants have been used experimentally to initiate flower buds on hydrangeas (21). Floral initiation by growth regulators: 1) eliminates the need for photoperiod control during high night temperature periods; 2) eliminates heat delay, bud abortion, and foliar distortion associated with high temperatures, as experienced under blackcloth; and 3) controls plant height. A-Rest®, Bonzi®, and Sumagic® all appear effective in stimulating hydrangea floral initiation. Undesirable side effects of using these chemicals for floral initiation are an increase in the forcing time and a reduction in plant height and inflorescence diameter upon forcing (Bailey, unpublished data). Further research is needed prior to commercial use, but results look promising, especially for growers in warmer areas, not experiencing 65°F night temperatures for floral initiation.

Defoliation and Dormancy

After floral initiation has been completed, hydrangeas require a period of cold storage to overcome bud dormancy, thus allowing rapid expansion of the inflorescences during forcing. Plants should be defoliated prior to placement into the cooler to help prevent the occurence of diseases, especially botrytis bud rot (see section "Pests, Diseases, and Other Problems").

Defoliation

Many methods of defoliation have been employed and none are without problems. The grower should experiment to find the method that fits his system best. The more common defoliation methods are described below and in Table 11.

The most effective chemical for hydrangea defoliation is 2-butyne-1, 4-diol. Butyne diol is a water soluble, brown crystal manufactured by GAF Corporation and Eastman Kodak Company. It is poisonous; harmful if inhaled or absorbed through the skin, and can be fatal if swallowed. Extreme caution should be observed when using butyne diol. Foliar sprays of 7500–12,500 ppm result in complete defoliation within 7–10 days (59). Complete spray coverage of the foliage is essential for best results; apply enough volume to wet all leaves (Table 11). Butyne diol sprays do not require an air-tight storage facility.

Table 11. **Defoliation spray solutions for hydrangeas.** Apply enough volume to wet all foliage thoroughly (0.75 to 1.0 gal of finished solution per 100 ft^2 of bed or bench area).

Chemical	Spray solution (ppm)	Amount/gallon of final solution	Amount/liter of final solution
BUTYNE DIOL	5000	0.67 oz/gal	5.05 g/liter
(990 g a.i./kg)	10,000	1.35 oz/gal	10.10 g/liter
	15,000	2.02 oz/gal	15.15 g/liter
ETHREL®	1000	0.53 fl oz/gal	4.2 ml/liter
(239.653 g	2000	1.07 fl oz/gal	8.3 ml/liter
a.i./liter)	3000	1.60 fl oz/gal	12.5 ml/liter
FLOREL®	1000	3.24 fl oz/gal	25.3 ml/liter
(39.543 g	2000	6.47 fl oz/gal	50.6 ml/liter
a.i./liter)	3000	9.71 fl oz/gal	75.9 ml/liter
PRO GIBB® 4%[a]	50	0.20 fl oz/gal	1.56 ml/liter
(32.123 g a.i./liter)			

[a]Gibberellic acid will not cause defoliation when used alone. It should be used as a defoliant enhancer to speed the leaf drop process in conjunction with a defoliating treatment. Use of gibberellic acid to assist defoliation can cause taller plants during forcing and reduce forcing time.

Plants can be treated while still in the greenhouse or outdoor growing area. Butyne diol is most effective between 60–70°F.

Although not an approved use of the chemical, Vapam® effectively defoliates hydrangeas if done properly (34). Vapam® requires an air-tight storage unit, with few leaks for best results. Use 10 teaspoons Vapam®/1000 ft^3 of building volume. Mix the Vapam® with enough water to evenly distribute over the walkway in the storage facility. If the storage building is not air-tight, a second treatment may be required 3–4 days later. Vapam® is most effective when temperatures are maintained between 60–70°F.

Over-treatment with Vapam® can injure plants and will result in poor foliage and flowering during forcing. People cannot enter the defoliation facility for at least 24 hours after treatment.

Ethylene is also used to defoliate hydrangeas. Ethylene gas can be purchased in pressurized cylinders and injected by bubbling through water into a 60°F storage room (59). Inject 1 ft³ of ethylene per 1000 ft³ of storage area over the course of 1 week. Larger volumes are required for leaky storage buildings.

Apples are an alternative source of ethylene that are still used (52). Place 1 bushel of ripe apples per 400 ft³ of air-tight building area. Keep the storage facility at 70°F during defoliation and use a fan to circulate the air within the room. Defoliation should be complete within 4–7 days.

The ethylene-releasing compound ethephon (Ethrel® or Florel®) causes hydrangea defoliation, but retards growth during forcing, even at the low dose of 1000 ppm (57,73). Reductions in inflorescence size are undesirable, so ethephon is not the best choice for defoliation as plants will be shorter during forcing and have smaller inflorescence diameters. Use 1000–3000 ppm a.i. as a foliar spray (Table 11). Apply sufficient spray to wet all foliage. Ethephon defoliation is best between 60–70°F.

Another potential defoliating agent is the organophosphorus compound tributyl phosphorotri-thioite (Folex®) currently used for cotton defoliation. Investigations conducted in the 1950s reported effective defoliation in 8 days with a 1–2% spray solution of Folex® (31). However, more work is needed to compare Folex® and other organophosphate sprays with defoliation treatments currently used on hydrangeas.

Shanks (59) recommends spraying plants with 50 ppm gibberellic acid (GA₃) to speed the defoliation process. The GA₃ (Pro Gibb 4%) can be incorporated into the defoliation spray if butyne diol, Florel®, or Ethrel® are

used or it can be sprayed prior to treatment with Vapam®
or ethylene gas (Table 11). The addition of 50 ppm GA_3
both hastens defoliation and enhances forcing if Ethrel® or
Florel® are used for defoliation (59).

Remove all abscised leaves from the pots prior to
placement into cold storage, as dead tissue encourages
botrytis bud rot during storage. After defoliation is complete, spray plants with benomyl (Benlate®) or chlorothalonil (Daconil®) to help prevent botrytis during
storage.

Cold Storage

Hydrangea dormancy requires a period with
temperatures below 55°F prior to bud break (62). Place
plants into a dark cooler set at 40–45°F for a minimum of 6
and a maximum of 8 weeks (34). If a storage period of
greater than 8 weeks is anticipated, store plants at 33–35°F
throughout the storage period. This lower temperature
helps prevent the excess stretching and shortened forcing
time associated with overly long storage at 40–45°F (34).
Shanks (59) reports better results (more uniform development during forcing) if a 52°F storage temperature is used
for a period of 8 weeks. The warmer temperature is especially useful if floral initiation is not complete at the time of
storage.

Many growers do not have dark cold storage facili-
ties so must rely on natural cooling to meet the hydrangea
dormancy requirement. Natural cooling requires protec-
tion from hard freezes as well as good ventilation for high
temperature control. Covered storage houses should be
maintained at 35°F minimum and 50°F maximum for best
results. For prolonged natural cooling (longer than 10
weeks) where plants receive light, buds may begin to
break while still in the storage house so the total forcing
time required is reduced.

40

Do not allow pots to dry out during storage or the plants could be damaged. Allowing the soil to dry out desiccates the root system and leads to poor root growth during forcing. Protect plants from direct air flow while in cold storage as constant air flow drys and damages buds. Maintain a relative humidity of 20–60%. Lower than 20% RH desiccates buds while humidities above 60% encourages botrytis and other rot organisms.

Greenhouse Forcing

Forcing is the final production phase. Total forcing time is cultivar and temperature dependant (See Table 1).

Transplanting

Hydrangeas are usually produced in 4–5 in. pots during the vegetative, floral initiation, and cold storage phases of production. The smaller root ball size is desirable to reduce shipping costs when the plants are transported to the forcer. Some producers ship dormant plants bare-root to forcers. If dormant plants are received bare-root, forcers should place plants into a pot the same size as the soil ball at the onset of forcing. During forcing, plants should be transplanted into the final pot size (See Table 8). Wait until roots are actively growing prior to transplanting. New root growth should be evident 2–4 weeks from the start of forcing. Transplanting prior to active root growth increases the incidence of root rot and causes leaf chlorosis (30).

Forcing Pink or White Plants

The sepal color program initiated during summer production must be maintained throughout the forcing period (see *Control of Sepal Color* in section "Plant Establishment and Vegetative Growth"). When transplanting pink hydrangeas into their final pot size, use a growth medium recommended for pink flowers (Table 3). Follow the fertilizer program given in Table 12. If an alternate plan is used, maintain high levels of nitrogen and phosphorus to assure pink coloration.

If raising media pH above 6.0 is difficult, increase the amount of phosphorus, as the additional phosphate depresses aluminum availability and allows for a clearer, pink flower color. Incorporate 4.5 lbs treble superphosphate/yd^3 of media instead of 3 lbs (see Table 3) and replace the liquid fertilizer selected from Table 12 with mono-ammonium phosphate (11-53-00) plus potassium chloride (00-00-60) every third feeding. Use 12.1 oz 11-53-00 plus 2.7 oz 00-00-60 per 100 gallons to substitute for continuous fertilization or use 28.9 oz 11-53-00 plus 5.5 oz 00-00-60 to substitute for intermittent fertilization. This extra phosphorus program is only necessary when the growth media pH is less than 6.2. If pH control is adequate, the normal fertilization programs outlined in Table 12 are sufficient during forcing.

Iron deficiency chlorosis can occur, especially with pink flowering plants when pH is greater than 6.2. Apply iron chelate drenches to control the problem (see *Nutrition and Fertilization* in section "Plant Establishment and Vegetative Growth" for iron chelate drench specifics).

Forcing Blue Plants

As with pink flowers, continuation of the sepal color program is necessary for forcing blue-flowering plants. Use the growth medium recommended for blue flowers (Table 3) when transplanting.

Blue-flowering plants require ample aluminum and a low pH (5.0–5.5) to make aluminum available. Apply 8 fl oz per 6 in. pot of aluminum sulfate as a soil drench (15

Table 12. **Liquid fertilization programs for forcing flowering.**

Pink or White Flowers

Continual fertilization. Supply 175–250 ppm N at each irrigation using:
  a. 9.3–13.4 oz 25-10-10 per 100 gallons *or*
  b. 2–2.5 oz 11-53-00 (monoammonium phosphate) plus 2–3 oz 13-00-44 (potassium nitrate) plus 6–8.4 oz 32-00-00 (Uran 32) per 100 gallons

Maintain irrigation solution pH at 6.3 using phosphoric, sulfuric, or citric acid. Phosphoric acid is preferred to maintain desired levels of P for pink flowers.

Intermittent fertilization. Supply 450–600 ppm N every 7–10 days using:
  a. 24–32 oz 25-10-10 per 100 gallons *or*
  b. 4.5–6.1 oz 11-53-00 (monoammonium phosphate) plus 5.5–7.3 oz 13-00-44 (potassium nitrate) plus 15–20 oz 32-00-00 (Uran 32) per 100 gallons

Maintain fertilizer solution and clear water pH at 6.3 using phosphoric, sulfuric, or citric acid. Phosphoric acid is preferred to maintain desired levels of P for pink flowers.

Blue Flowers

Continual fertilization. Supply 150–200 ppm N & 300–350 ppm K at each irrigation using:
  2.7–4.8 oz 21-00-00 (ammonium sulfate) plus 11–12.8 oz 13-00-44 (potassium nitrate) per 100 gallons

Maintain irrigation solution pH at 5.3 using sulfuric or citric acid. Do *not* use phosphoric acid; the added P is counter-productive.

Intermittent fertilization. Supply 400–550 ppm N & 500–600 ppm K every 7–10 days using:
  14.1–21.4 oz 21-00-00 (ammonium sulfate) plus 18.3–21.9 oz 13-00-44 (potassium nitrate) per 100 gallons

Maintain fertilizer solution and clear water pH at 5.3 using sulfuric or citric acid. Do *not* use phosphoric acid; the added P is counter-productive.

lbs/100 gallons) to pre-moistened soil (59) as the root system may be burned if the soil is dry. Keep foliage dry to prevent leaf burn. Make the first application after plants are well established, 7–10 days after transplanting. If 5.0–5.5 pH medium and irrigation solution is maintained, 2 applications are sufficient. Additional applications may be necessary if the irrigation solution is not acidified or if the growth media pH is not sufficiently low. Measure soil pH 10–14 days after each drench. If the pH is higher than 5.5, another application is necessary.

Blue-flowering plants receiving adequate phosphorus (P) fertilization prior to cold storage do not need P during forcing. High levels of P during forcing result in unclear blue or mauve color. Maintain high levels of potassium (K) to assure clear blue coloration. The fertilizer program in Table 12 is designed to supply ample K. If using an alternate fertilizer program, provide levels of K similar to those outlined in Table 12.

Watering

When plants are removed from cold storage, grow them on the "dry side" to encourage root development (30). However, frequent overhead syringing should be employed to help stimulate shoot growth. Discontinue syringing once buds break and new leaves begin to expand.

Do not begin fertilizing plants until root activity is obvious. Once root growth occurs and plants have been transplanted to the final pot size, maintain uniform soil moisture and do not allow plants to wilt at any time. Maintain the appropriate water and fertilizer solution pH for the flower color being grown (see Table 12). See *Watering* in section "Plant Establishment and Vegetative Growth" for a more detailed discussion of watering and water quality.

Light Intensity and Temperature

Plants should be given full sun (up to 7500 ft-c) up to the time of flower color during the darker months of forcing. Plants should be spaced far enough apart to prevent mutual shading. Poor light leads to "soft" growth and prevents plants from reaching their full potential. Final spacings for single stem, 2-stem, 3-stem, and 4-stem plants are 8 × 8 in., 12 × 12 in., 13 × 14 in., and 14 × 15 in., respectively. Once flowers show color, shading to 3000 ft-c is beneficial to prevent petal burn from bright light and desiccation. Earlier shading may be needed by growers located in more southerly latitudes, especially if using late-season forcing.

The rate of plant development during forcing is directly related to temperature. Generally, plants are forced at a 60°F night temperature/75°F vent temperature until flowers begin to show color (about 2½ weeks prior to sale) then temperatures are dropped to 54°F night/65°F vent to intensify flower color. However, depending on the desired sales date, temperature can be adjusted to speed up or slow down the forcing (Table 13).

Temperature also affects plant habit. Cooler forcing temperatures (54°F nights throughout the entire forcing period) lead to longer internodes, larger leaves, taller plants, and larger-diameter inflorescences than plants forced at greater than 60°F (36,59). Regardless of the forcing temperature used prior to flower color, all plants should be finished at 54°F nights for best flower color and post-production life.

An alternative to a continuous night temperature is a split-night temperature, utilizing a lower temperature between 2–8 AM. Shanks (60) reports that a 63°F night temperature with 54°F between 2–8 AM (75°F venting temperature) resulted in an identical forcing time but larger inflorescences compared to a constant 63°F night temperature. The advantage of a split-night temperature scheme is a reduction in heating costs during the winter months and should be considered by forcers in the northerly latitudes.

Table 13. Temperature effects on forcing schedules. The table is based on 'Merritt's Supreme' and other cultivars that bloom in 88 days using 60°F night temperatures. The timings given assume a night temperature of 54°F the last 18 days of forcing. Adapted from (34, 59, 86).

Time interval	Night temperature in °F (°C)		
	54 (12)	60 (15)	66 (19)
Days from start of forcing to bloom	112	88	80
Days from start of forcing to pea-sized inflorescence (3/16 in., 5 mm diameter)	42	32	28
Days from pea-sized (3/16 in. or 5 mm) to bloom	70	56	52
Days from U.S. nickel-sized inflorescence (13/16 in., 21 mm diameter) to bloom	53	42	39
Days from U.S. silver dollar-sized inflorescence (1½ in., 38 mm diameter) to bloom	35	28	26
Days from first color (drop night temperature to 54°F) to bloom	18	18	18

Photoperiod

If plants have been sufficiently cooled (at least 6 weeks at 40–45°F), photoperiod manipulation is not necessary and is not beneficial (68). However, if inflorescence primordia were not adequately developed prior to cold storage, or if plants do not receive sufficient storage prior to forcing, long days should be provided during forcing. For example, for plants receiving only 7 weeks of cooling at 52°F, a 4 hour night break (lights on from 10:00 PM–2:00 AM) reduced forcing time by 10%, increased height by 20%, and increased inflorescence diameter by 10% compared to plants which received 12 hour photoperiods, similar to natural daylength during forcing (63). Growers should consider giving an "insurance policy" 4 hour night break during forcing if insufficient cooling is suspected or if warm temperatures were used during storage.

Chemical Growth Retardants

Hydrangeas require height control during forcing, especially if low forcing temperatures are used. Sprays of B-Nine® are commonly used at 2500 ppm (most cultivars) to 5000 ppm (for tall growing cultivars such as 'Rose Supreme') (See Table 9). Spray plants after 3–5 leaf pairs have unfolded, about 3 weeks from the start of forcing using a 60°F night temperature. If low forcing temperatures (below 60°F) are used or if light intensity is low, multiple applications may be needed. Also, if plants have received excessively long cold storage (greater than 8 weeks at 40–45°F or greater than 10 weeks of natural cooling), plants tend to grow taller so require more height control. Repeat applications every 10–14 days as needed. Do not apply growth retardants after flower buds are larger than ¾ in. in diameter or inflorescences will be reduced in size.

Other chemical growth retardants are effective for hydrangea height control. (See *Chemical Growth Retardants* in section "Plant Establishment and Vegetative Growth" for descriptions and Tables 9 and 10 for recipes.)

GA$_3$ Treatments

The use of GA$_3$ sprays as a substitute for cold storage has met with mixed success (18,67). It is possible to flower hydrangeas without defoliation and cold storage if GA$_3$ is sprayed onto flower-budded plants (18,59). However, plant quality is not as high as when receiving traditional cold storage treatment, unless the plant is being grown as a "mini-pot" hydrangea (see section "Summarization of Production Schedules" for mini-pot schedules). Sprays of GA$_3$ are useful for traditional production if plants have not received adequate cold storage and are slow to develop or dwarfed. If GA$_3$ sprays are used, apply at 5 ppm (59). Multiple applications may be needed to overcome insufficient cold storage. Growth must be carefully monitored to prevent excessive stretching due to GA$_3$. Discontinue the sprays as soon as normal growth (normal-sized leaves and internodes) has resumed. Application of GA$_3$ can be viewed as another "insurance policy", similar to long days during forcing. In most cases, it is not needed.

Hardening Plants

Hydrangeas require acclimation prior to sale for best flower color and post-production performance. As sepal color becomes evident (about 2½ weeks prior to sale), reduce fertilization by ½ to prepare plants for the post-production environment. If forcing at a night temperature warmer than 54°F, lower the night temperature to 54°F and reduce the venting temperature to 65°F as color appears on the inflorescences. Light shading may be required to prevent burning of petals as coloration progresses. Reduce the amount of watering as much as possible during the last 3 weeks of production; however, under no circumstances should plants be allowed to wilt.

Pests, Diseases, and Other Problems

Hydrangea problems vary with location, production phase, and production system. For example, the most severe pests during outdoor vegetative growth may be slugs and snails, while during greenhouse forcing two-spotted mites are most common.

The best pest management program prevents the occurrence of problems rather than trying to remedy problems once they appear. If 20% of the plants in cold storage show signs of botrytis bud rot, the grower has already suffered tremendous losses. Good production techniques must be incorporated into the pest management program. For example: 1) steam pasteurize growth media and containers prior to use; 2) do not allow weeds to invade growing areas; 3) dispose of unnecessary plant materials in the growing area; 4) examine plants continuously for signs and/or symptoms of pests and diseases. Every cultural practice precluding the appearance of pathogens and pests should be aggressively used. No pest nor pathogen has yet become resistant to good cultural management.

Unfortunately, even exceptional cultural practices may occasionally prove insufficient, so some chemical control must be utilized. Whenever possible, keep chemical usage to a minimum, and rotate chemicals to reduce pest resistance.

Chemical regulations differ from jurisdiction to jurisdiction and furthermore are constantly changing. Before purchasing and using any pesticide, check all labels for registered use, concentrations, and frequency of application together with safety precautions. It is illegal to use any pesticide or growth regulator inconsistent with the current labelling.

Table 14 lists hydrangea pests and diseases. Of the pests and diseases listed, the most common and potentially severe are discussed below. The chemicals listed in this chapter are presented for *information only*. No endorsement is implied by mention, nor is criticism suggested by non-mention of a product.

Table 14. **Potential pests and diseases of hydrangeas.**
Adapted from (35, 50, 72, 86).

Pests

Insects
  Aphids (*Aphis gossypii, Myzus circumflexus, M. persicae*)
  Four-lined plant bug (*Poecilocapsis lineatus*)
  Leaf-tiers (*Exartema ferriferanum, Udea rubigalis*)
  Rose-chafer (*Macrodactylus subspinosus*)
  Scale (*Lepidosaphes ulmi, Pulvinaria* spp.)
  Tarnished plant bug (*Lygus lineolaris*)
  Thrips (*Hercinothrips femoralis*)
  Whiteflies (*Bemisia tabaci, Trialeurodes vaporariorum*)

Mites
  Two-spotted mite or Red spider mite (*Tetranychus urticae*)

Other pests
  Slugs (*Deroceras reficulatum, Limax* spp.)
  Snails (*Helix* spp.)

Diseases

Bacteria
  Bacterial Wilt (*Pseudomonas solanacearum*)

Fungi
  Blister rust (*Pucciniastrum hydrangeae*)
  Bud rot (*Botrytis cinerea*)
  Gray mold (*Botrytis cinerea*)
  Inflorescence blight (*Botrytis cinerea*)

52

Leaf spots (*Ascochyta hydrangeae, Cerospora arborescentis, Corynespora cassicola, Phyllosticta hydrangeae, Septoria hydrangeae*)

Powdery mildew (*Erysiphe polygoni*)

Root rot (*Armillaria* spp., *Polyporus* spp., *Rhizoctonia* spp., *Sclerotium* spp.)

Stem rot (*Polyporus* spp., *Rhizoctonia* spp., *Sclerotium* spp.)

Mycoplasma-Like Organisms (MLO)

Hydrangea virescence

Nematodes

Leaf nematodes (*Aphelenchoides* spp.)

Lesion nematodes (*Pratylenchus* spp.)

Root-knot nematodes (*Meloidogyne incognita, M. hapla*)

Stem nematodes (*Ditylenchus dipsaci*)

Physiological Disorders

Hydrangea distortion

Iron deficiency chlorosis

Viruses

Alfalfa mosaic virus

Cucumber mosaic virus

Hydrangea mosaic virus

Hydrangea ring-spot virus

Tobacco necrosis virus

Tobacco rattle virus

Tobacco ring-spot virus

Tomato ring-spot virus

Tomato spotted-wilt virus

Pests

APHIDS

Aphids are a problem during outdoor vegetative growth, greenhouse vegetative growth, and greenhouse forcing stages. Not only do aphids cause damage through feeding, they also are vectors for viruses.

Biological control: Aphids in the greenhouse may be controlled with the predatory midge, *Aphidoletes aphidimyza*, while "aphid lions", *Chrysopa carnea*, may be effective for outdoor control (42). Growers interested in implementing an integrated pest management program involving biological controls are encouraged to read the articles by Aylsworth (11), Miller (42), and van Lenteren and Woets (85).

Chemical control: In conjunction with cultural and biological control measures, aphids may be controlled with acephate (Orthene® 75SP) (6–16 oz/100 gal; outdoor use only), bifenthrin (Talstar® 10WP) (6–24 oz/100 gal), chlorpyrifos (Dursban® 50WP) (8–16 oz/100 gal), dichlorvos (Vapona®) (smoke canister), endosulfan (Thiodan® 50WP) (16 oz/100 gal), or fluvalinate (Mavrik Aquaflow®) (2–5 fl oz/100 gal).

TWO-SPOTTED SPIDER MITES

Sometimes called red spiders, two-spotted mites can rapidly become a problem during greenhouse production of hydrangeas. Removal or isolation of highly infected plants helps prevent spreading.

Biological control: The predatory mites *Phytoseiulus persimilis, Phytoseiulus longipes, Amblyseius californicus,* and *Metaseiulus occidentalis* are useful for mite control (42).

Chemical control: Miticides such as abamectin (Avid® 0.15EC) (8 fl oz/100 gal), bifenthrin (Talstar® 10WP) (6–25 oz/100 gal), dicofol (Kelthane® 35WP) (16–24 oz/100 gal), dienochlor (Pentac Aquaflow®) (8 fl oz/100 gal), fenbutatin-oxide (Vendex® 50WP) (4–16 oz/100 gal), fluvalinate (Mavrik Aquaflow®) (4–5 fl oz/100 gal), or propargite (Omite® 30W) (1–2.5 lb/100 gal) may be effective. Since mites prefer to feed on the undersides of leaves, complete coverage is essential in chemical control.

SLUGS AND SNAILS

These mollusks can be very damaging in an outdoor production setting. Growing on a clean bed or plastic-covered growing area helps prevent infestation.

Chemical control: Use a mollusk-specific pesticide such as metaldehyde (Slugit®) (1 fl oz/gal) or methiocarb (Slug-Geta® Pellets) (1 lb of bait/1000 ft^2).

WHITEFLIES

Although hydrangeas are not first choice for whiteflies, infestations can occur during greenhouse production. Elimination of unnecessary plants in and around the growing area helps prevent whitefly populations prior to hydrangea forcing.

Biological control: The chalcidoid wasp, *Encarsia formosa* (11) and the whitefly-parasitic fungi *Aschersonia aleyrodis* and *Verticillium lecanii* (85).

Chemical control: Use an insecticide such as acephate (Orthene® 75SP) (6–16 oz/100 gal; outdoor use only), bifenthrin (Talstar® 10WP) (6–24 oz/100 gal), dichlorvos (Vapona®) (smoke canister), endosulfan (Thiodan® 50WP) (16 oz/100 gal), fluvalinate (Mavrik Aquaflow®) (2–5 fl oz/100 gal), or resmethrin (26EC) (16 fl oz/100 gal).

The following pesticides have been reported to cause injury on hydrangeas: chlorobenzilate, demeton, diazinon, dimethoate (Cygon 2-E®), malathion EC, nicotine sulfate (smoke), oxythioquinox (Morestan®), parathion WP, and trichlorfon (22).

Diseases

Although many diseases are listed in Table 14, only botrytis and powdery mildew are widespread and severe. Good cultural practices and proper water management should prevent root rot and leaf spot diseases. As with most plants, maintaining dry foliage is essential for an adequate disease prevention program.

BUD ROT, GRAY MOLD, INFLORESCENCE BLIGHT, AND STEM ROT

By far the most damaging pathogen affecting hydrangeas is *Botrytis cinerea.* Under humid conditions during propagation, gray mold can appear on leaves and stems. Bud and stem rot can occur during cold storage if not prevented. Finally, inflorescence blight can occur during forcing if the inflorescences are inadvertently watered overhead. Controlling humidity and maintaining proper plant spacing may help deter *Botrytis,* but chemical prevention is usually required. Cultivars differ in their susceptibility to *Botrytis.* Powell (53) gives the following list for most susceptible to least susceptible: 'Merveille' (most susceptible), 'Rose Supreme', 'Todi', 'Improved Merveille', 'Regula', 'Meritt's Supreme', 'Sister Therese', 'Strafford', and 'Kuhnert' (least susceptible).

Chemical control: Fungicides include benomyl (Benlate® 50DF) (0.25–4 lb/100 gal), chlorothalonil (Daconil 2787® 75WP) (1.5 lb/100 gal), dicloran (Botran® 75WP; outdoor use only) (1–5 lb/100 gal), iprodione (Chipco 26019® 50WP; Rovral®), (1–2 lb/100 gal), mancozeb (Dithane M-45® 80WP; outdoor use only) (24 oz/100 gal), and vinclozolin (Ornalin® 50WP) (16–24 oz/100 gal).

POWDERY MILDEW

Also preferring high humidity conditions, powdery mildew can become serious during outdoor and greenhouse production.

Chemical control: Effective fungicides include benomyl (Benlate® 50DF) (0.25–4 lb/100 gal), dinocap (Karathane® 19.5WP) (8 oz/100 gal), fenarimol (Rubigan® 12.5EC; outdoor use only) (3 fl oz/100 gal), and sulfur burn pots.

HYDRANGEA VIRESCENCE

Also referred to as phyllody, this disease is usually

observed as a greening of the inflorescence, yellowing of leaves, and stunting of new growth (35). The incidence of hydrangea virescence, caused by mycoplasma-like organisms (MLO), has declined as heat treatment and tissue culture have been utilized by propagators to produce pathogen-free stock plants.

HYDRANGEA DISTORTION

For years, hydrangea producers in warm areas have encountered distorted foliage during high temperature periods in the summer (Fig. 4). Weiler and Lopes (87) first described this problem and named it "hydrangea distortion." Affected plants develop thickened leaves that are narrow and sometimes mottled. Evidence suggests the cause of hydrangea distortion is environmental stress (temperatures greater than 86°F for extended periods of time) and not a virus, MLO, or other pest organism (13). When temperatures moderate, plants resume normal growth and normal leaves develop. If normal growth resumes prior to the autumn floral initiation period, inflorescences will develop normally. However, late "heat waves" during floral initiation result in distorted inflorescences upon forcing. Cultural practices including shading, syringing, and utilizing fan and pad cooling reduce leaf temperature and minimize the likelihood of hydrangea distortion.

IRON DEFICIENCY CHLOROSIS

This problem can occur from propagation to flowering and is a result of high pH in the growth medium. Interveinal chlorosis starts with the young, developing leaves and can become quite severe, if not corrected (Fig. 5). Reduce the growth media pH to below 6.5 and apply iron chelate at first signs of chlorosis (see section "Plant Establishment and Vegetative Growth" for chelate drench recipe).

Post-Production Care

Growers can help assure optimal post-production life span and hence customer satisfaction by hardening plants (outlined in section "Greenhouse Forcing"). Hydrangeas have the durability of chrysanthemums in terms of post-production life span if consumers are familiar with one word: water. Once color shows in a hydrangea inflorescence, the plant never fully recovers from wilt. Attempts to slow water loss of hydrangeas with antitranspirants have been unsuccessful (75), so consumers must rely on maintaining a constant supply of moisture to plants. Hydrangeas will survive longer at cool (60–65°F) indoor temperatures than at warm (72–78°F) temperatures. Filtered sunlight is better than direct light or dense shade. Most hydrangea cultivars are winter-hardy to 0°F, USDA zone 7a (37). Consumers living in zones 7a–10a should be able to successfully transplant hydrangeas to their yard, if adequate soil and cultural conditions are met.

Summarization of Production Schedules

Traditional Production Schedules (Easter to Mother's Day)

Table 8 outlines summer culture schedules for each cutting type, assuming a cold storage starting date of late October-early November. Follow this table for propagation and pinching during the summer vegetative growth phase for traditional Easter-to-Mother's Day sales.

Place plants into cold storage only after adequate inflorescence initiation has occurred and when the inflorescence is in stage 5 or 6 (Figure 2). If using 40–45°F cold storage, store plants for a minimum of 6 weeks and a maximum of 8 weeks. If using 52°F storage, store plants for a minimum of 8 weeks. Do not store plants at 52°F longer than 8 weeks or premature bud growth in the cooler may occur. If more than 8 weeks of cold storage is needed for the scheduled harvest date, use 33–35°F cold storage instead of 40–45°F or 52°F.

Forcing time requirements depends on cultivar (Table 1) and forcing temperatures (Table 13). Using different cultivars and different forcing temperatures, it is possible to schedule flowering throughout the spring months (Table 15).

Valentine's Day Schedules

Shanks (59) reports the possibility of forcing hydrangeas into bloom for Valentine's Day. Two schedules for Valentine's Day are given in Table 15. Notice the differences in propagation dates, defoliation methods, cold storage dates, cold storage temperatures, forcing time required, and forcing conditions required. Shanks also recommends 3 applications of B-Nine® during the summer vegetative growth phase of a Valentine's Day crop: June 15, July 15, and August 15, using 5000 to 7500 ppm B-Nine®. These schedules fit nicely into a poinsettia growing scheme that uses an early (mid-November) poinsettia sales date.

Year-Round Production of Mini-Pot Hydrangeas

The concept of forcing hydrangeas year-round has been thoroughly investigated (12,58). The methods employed allow for vegetative growth to attain plant size, then rapid floral initiation and inflorescence expansion. Growers interested in producing mini-pot hydrangeas should select medium-to-small sized rapidly forcing cultivars such as 'Merritt's Supreme' and 'Dr. Bernard Steiniger' (Figure 6).

To obtain vegetative cuttings, follow the points outlined under *Maintaining Stock Plants* in section "Propagation." Propagate shoot tip cuttings under continuous photoperiod following guidelines outlined in section "Propagation." Rooting should be sufficiently completed for transplant to a 4½–5½ in. pot in 4 weeks. After transplant, grow plants for 2 weeks at 62°F minimum temperature under continuous photoperiod.

Floral initiation is stimulated by cool temperatures and short photoperiods. Do not defoliate plants for mini-pot production. If natural conditions are accommodating, stimulate floral initiation using 8 hour photoperiods in a 52°F night temperature/62° day temperature greenhouse for 6 weeks. If ambient temperatures are too high, place plants in a cooler set at a constant 52°F. Light for 12 hours daily to retain foliage, supplying 100 ft-c at the plant level (59). Continue for a total of 6 weeks.

After the floral initiation phase, force plants in a 62°F minimum temperature greenhouse (up to 72°F minimum temperature is acceptable during the summer months). Supply a 4 hour night break (10 ft-c incandescent light from 10:00 PM–2:00 AM) during forcing. Spray GA$_3$ at 25–50 ppm during the first week of forcing to encourage rapid inflorescence development. A second spray may be needed 2 weeks later, if the buds do not begin to expand. The forcing period is dependant on the time of year, ranging from 8 weeks during the warmer and brighter conditions of the summer months up to 16 weeks during the cooler and darker conditions of the winter months.

Table 15 outlines two sample schedules for mini-pot hydrangeas. These schedules are given as a starting point for growers to build upon. More work is needed to "fine-tune" production techniques for different climatic conditions. Also, available cultivars need to be screened for adaptability to mini-pot production techniques.

Table 15.		**Hydrangea production schedules.**

Days prior to sale	*Date*	*Procedure*

A. Easter with sales date of April 8.

'Todi' or 'Merritt's Supreme', 3–4 inflorescences per plant

335	05/08	Stick shoot tip cuttings
286	06/26	Pinch plants
160	10/30	Inflorescences should be at Stage 5–6; apply defoliation treatment
153	11/06	Place defoliated plants into 40–45°F dark storage
111	12/18	Place plants in 54°F NT greenhouse to begin forcing
96	01/02	Root growth should be evident; transplant into 7–8 in. pots
69	01/29	Inflorescences should be visible (pea-size); if not, increase NT to 58–60°F
33	03/06	Inflorescences should be 1½ in. in diameter; if not, adjust NT accordingly
18	03/21	Color should be present on inflorescences; adjust NT to 54°F for color intensification
0	04/08	Sales date

'Kasteln' or 'Schenkenburg', 1 inflorescence per plant

300	06/12	Stick shoot tip cuttings
160	10/30	Inflorescences should be at Stage 5–6; apply defoliation treatment
153	11/06	Place defoliated plants into 40–45°F dark storage
95	01/03	Place plants in 60°F NT greenhouse to begin forcing
81	01/17	Root growth should be evident; transplant into 5½–6 in. pots
75	02/04	Inflorescences should be visible (pea-size); if not, adjust NT accordingly
28	03/11	Inflorescences should be 1½ in. in diameter; if not, adjust NT accordingly
18	03/21	Color should be present on inflorescences; adjust NT to 54°F for color intensification
0	04/08	Sales date

Days prior to sale	Date	Procedure

B. Mother's Day with sales date of May 6.

'Rose Supreme', 3–4 inflorescences per plant

363	05/08	Stick shoot tip cuttings under mist
314	06/26	Pinch plants
188	10/30	Inflorescences should be at Stage 5–6; apply defoliation treatment
181	11/06	Place defoliated plants into 33–35°F dark storage
102	01/24	Place plants in 60°F NT greenhouse to begin forcing
81	02/14	Root growth should be evident; transplant into 8 in. pots
62	03/05	Inflorescences should be visible (pea-size); if not, adjust NT accordingly
31	04/05	Inflorescences should be 1½ in. in diameter; if not, adjust NT accordingly
18	04/18	Color should be present on inflorescences; adjust NT to 54°F for color intensification
0	05/06	Sales date

C. Easter with sales date of March 24.

'Todi' or 'Merritt's Supreme', 3–4 inflorescences per plant

331	05/07	Stick shoot tip cuttings
282	06/25	Pinch plants
146	10/29	Inflorescences should be at Stage 5–6; apply defoliation treatment
139	11/05	Place defoliated plants into 40–45°F dark storage
88	12/26	Place plants in 60°F NT greenhouse to begin forcing
74	01/09	Root growth should be evident; transplant into 7–8 in. pots
56	01/27	Inflorescences should be visible (pea-size); if not, adjust NT accordingly
28	02/24	Inflorescences should be 1½ in. in diameter; if not, adjust NT accordingly
18	03/06	Color should be present on inflorescences; adjust NT to 54°F for color intensification
0	03/24	Sales date

Days prior to sale	Date	Procedure

D. Valentine's Day with sales date of February 7.[a]

'Todi' or 'Merritt's Supreme', 3–4 inflorescences per plant

303	04/10	Stick shoot tip cuttings
261	05/22	Pinch plants
156	09/04	Place un-defoliated plants into 52°F dark storage; allow natural defoliation, removing leaves from the floor as they abscise
100	10/30	Place plants into 62°F NT greenhouse to begin forcing; supply 10 ft-c incandescent light from 10:00 PM–2:00 AM during forcing
79	11/20	Root growth should be evident; transplant into 5½–6 in. pots
0	02/07	Sales date

'Todi' or 'Merritt's Supreme', 1 inflorescence per plant

275	05/08	Stick shoot tip cuttings
156	09/04	Place un-defoliated plants into 52°F dark storage; allow natural defoliation, removing leaves from the floor as they abscise
100	10/30	Place plants into 62°F NT greenhouse to begin forcing; supply 10 ft-c incandescent light from 10:00 PM–2:00 AM during forcing
79	11/20	Root growth should be evident; transplant into 5½–6 in. pots
0	02/07	Sales date

E. Mini-Pot Hydrangeas for Christmas with sales date of December 17.[a]

168	07/02	Stick shoot tip cuttings under a continuous photoperiod (CP); maintain 72°F bottom heat
140	07/30	Pot plants into 4½–5½ in. pots; allow vegetative growth under CP; maintain 62°F minimum temperature
126	08/13	Place plants into a 52°F cooler; supply 12 hours of 100 ft-c lighting daily

[a]Valentine's Day and mini-pot schedules adapted from (59).

66

Days prior to sale	Date	Procedure
84	09/24	Place plants into a 62°F night temperature greenhouse; Use 10 ft-c night break from 10:00 PM–2:00 AM; spray with 25–50 ppm GA_3; apply a second spray, if needed
0	12/17	Sales date

F. Mini-Pot hydrangeas for Canada Day (July 1) with sales date of June 25.[a]

154	01/22	Stick shoot tip cuttings under CP; maintain 72°F
126	02/19	Pot plants into 4½–5½ in. pots; allow vegetative growth under CP; maintain 62°F minimum temperature
112	03/05	Reduce temperatures to 52°F NT/62°F DT; supply 8 hour photoperiods using black cloth from 4:00 PM–8:00 AM
70	04/16	Increase temperatures to 62°F NT (higher NT is acceptable—up to 72°F)/72°F venting; use a 10 ft-c night break from 10:00 PM–2:00 AM; spray with 25–50 ppm GA_3
0	06/25	Sales date

Costs of Growing Hydrangeas

The many production alternatives reflected in overhead costs—outdoor lath house vegetative growth vs. greenhouse vegetative growth; winter fuel costs for heating and forcing temperature chosen; targeted holiday and the date in a given year, local taxes—prevent a single table from accurately reflecting hydrangea costs for every grower. However, cost accounting is necessary for all growers, and estimates of production costs should be made prior to production and sales. A guide to costs of forcing hydrangeas follows. This projection does not include costs of propagation, growing on, or cold storage.

Overhead (unallocated) Costs

Table 16 is a guide to estimate overhead (unallocated) costs for forcing hydrangeas. The table is based on the assumption that the greenhouse is in operation 52 weeks of the year with hydrangeas in a rotation with other potted flowering plants. The bottom line (line 16) indicates that space in this greenhouse costs approximately 15¢/ft^2/week.

Direct (allocated) Costs

Table 17 details allocated and total costs for forcing hydrangeas. The final column (column 12) indicates the cost of production per pot. This cost does not include profit, and profit should be included in the sales price. For example, for a desired profit of 12% on a two-bloom hydrangea, multiply the cost ($4.34) by 1.36. The result is the appropriate sales price, $4.93. A comparison between hydrangea product profitability is given in Table 18. From this type of table, a grower (assuming the market will support the product) can decide the best product to produce given known production costs. In the example, profit is maximized by producing single-bloom plants. These types of marketing decisions must take into account how many plants of each product the market will support. Notice also that the single-bloom production scheme has the highest direct cost figure (Table 17, col. 1), indicating a higher cash flow/short-term investment requirement so cash-on-hand may influence the product and profit.

Table 16.   Determining cost/ft^2/wk of forcing hydrangeas. Example assumes 30,000 ft^2 of growing space (40,000 ft^2 actual greenhouse space), and 30,000 two bloom 6 in. pots spaced at 12×12 in. throughout the 91 days of forcing. The example is carried forward to product line 2, **Table 17.** Adapted from (65).

Line	Item	Cost from greenhouse records	Less sum of costs directly related to hydrangeas	Overhead costs
1	Salaries, wages (Owner-manager plus employee wages and benefits)			90,000
2	Dormant 2-cane plants (including shipping, @ $1.90/plant	57,000	57,000	
3	6 in. azalea pots, @ 11¢/pot	3,300	3,300	
4	Growth medium, assuming 4 in. root ball on dormant plants; 32 yd^3, @ $65/yd^3	2,080	1,760	320
5	Fertilizer, assuming daily 12 fl oz/pot irrigations	1,350	1,050	300
6	Repairs, maintenance	4,000		4,000
7	Other direct costs (tags, pesticides, growth retardants, etc.)	1,962	1,762	200
8	*Total direct costs*		64,872	
9	Utilities	70,000		70,000

Line	Item	Cost from greenhouse records	Less sum of costs directly related to hydrangeas	Overhead costs
10	Administrative costs, marketing costs	5,000		5,000
11	Fixed over-head (insurance, depreciation, rent, taxes interest)	60,500		60,500
12	*Total unallocated costs*			230,320
13	Net growing area (ft^2)		30,000	
14	Unallocated costs/ft^2/year (12 ÷ 13)		7.68	
15	Weeks in operation		52	
16	*Unallocated costs/ft^2/wk (14 ÷ 15)*		0.1476	

Table 17.

Determining forcing costs for hydrangeas. Adapted from (65).

	Direct (allocated) costs			Overhead (unallocated) costs						Total cost per pot		
	Col 1	Col 2	Col 3	Col 4	Col 5	Col 6	Col 7	Col 8	Col 9	Col 10	Col 11	Col 12
Product	Sum of direct costs	No. of pots	Direct cost per unit	Space/pot (ft²)	Space/crop (ft²)	weeks	Space-weeks	Cost/ft²/wk	Indirect cost per pot	Cost/pot subtotal	Percent saleable	Adj cost/pot
Single bloom, 5½ in. pot	119,411	67,500	1.77	0.444	30,000	13	390,000	0.15	0.87	2.64	.95	2.78
Two-bloom, 6 in. pot; (from line 8, Table 16)	64,872	30,000	2.17	1.000	30,000	13	390,000	0.15	1.95	4.12	.95	4.34
Three-bloom, 6 in. pot	54,917	23,736	2.32	1.264	30,000	13	390,000	0.15	2.47	4.79	.95	5.05
Four-bloom, 7 in. pot	55,736	20,571	2.71	1.458	30,000	13	390,000	0.15	2.85	5.56	.95	5.86

Direct Costs

Column 1—Sum of costs directly assignable to the crop: pots, soil, fertilizer, labels, plant material, etc.
Column 2—Number of pots produced. Based on filling 30,000 ft² of bench area.
Column 3—Direct cost/pot. Col 1 ÷ by Col 2.

Overhead Costs

Column 4—Space/pot. Based on 8 × 8 in., 12 × 12 in., 13 × 14 in., and 14 × 15 in. for 1-, 2-, 3-, and 4-bloom plants, respectively.
Column 5—Space/crop. Col 2 × Col 4.
Column 6—Weeks during forcing.
Column 7—Time on bench × space occupied. Col 5 × Col 6.
Column 8—Cost/ft²/wk. Line 16, Table 16.
Column 9—Unallocated cost/pot. (Col 7 × Col 8) ÷ Col 2.

Total Costs

Column 10—Cost/pot subtotal. Col 3 + Col 9.
Column 11—Percent saleable either projected or from records.
Column 12—Actual cost/pot. Col 10 ÷ by Col 11.

Table 18. **Net profit based on costs and greenhouse conditions from Tables 16 and 17 with a 10% profit on sales.**

Product	No. of pots sold	Selling price/pot	Total sales	Total profit
Single bloom, 5½ in. pot	64,125	3.09	198,146	19,814
Two-bloom, 6 in. pot	28,500	4.83	137,655	13,765
Three-bloom, 6 in. pot	22,549	5.62	126,725	12,672
Four-bloom, 7 in. pot	19,542	6.52	127,413	12,741

The Future

Floriculture, In General

The floriculture industry is in the forefront of technology applications. We have experienced an ever increasing sophistication in greenhouses and production systems over the past 15 years. This trend should continue. Along with changes in technology, tighter regulations on chemical usage in agriculture will occur in the next 5–10 years. Growers may be forced to use recirculating "Ebb and Flow" benching systems to reduce the amount of pesticide and fertilizer residue reaching the ground water. Attitudes towards pest control must change, and biological controls and integrated pest management should be utilized more aggressively. Overall, the job of growing will become more demanding, requiring highly trained individuals with critical thinking skills.

Year-Round Production of Hydrangeas

The most exciting thing I see in the future of hydrangeas is the refinement of year-round production procedures based on increasingly more profound understandings of plant behavior. Manipulation of vegetative growth, floral initiation, and inflorescence expansion is now understood. The challenge now is to 1) document differences in scheduling during the course of the year and 2) select the most rapidly growing/responding cultivars. Different cultivars may be better suited for different harvest dates, in a case similar to chrysanthemums. For example, 'Rosa Rita' produces a nice single bloom plant during cool, low light periods but grows poorly during the summer. With over 500 cultivars available, cultivar screening at the local level is essential.

More research is needed to investigate the role of chemical growth retardants in floral initiation. The procedure is effective, but most techniques increase forcing time. Current studies are underway to use chemical growth retardants for floral initiation followed by GA_3 sprays to stimulate inflorescence expansion without any cold storage treatment. In the future, growers should be able to propagate vegetative plants, initiate inflorescences chemically, and produce flowering plants the year around.

Breeding

Hydrangea breeding has traditionally been a European undertaking, so most North American cultivars have derived from European breeding programs. In many cases, traits North American growers require are not a priority in European programs.

For example, hydrangea distortion is a significant and widespread problem in North America. It is rarely a problem in the more moderate summers of northern Europe and has not been addressed in any breeding program on that continent. Unfortunately, the cultivars most susceptible to the disorder ('Rose Supreme', 'Todi', and 'Merritt's Supreme') are the most popular cultivars grown in North America. Cultivars, however, vary in their tolerance of high temperature conditions. For example, 'Blau Donau' plants appear resistant to hydrangea distortion. Breeding for resistance to hydrangea distortion is needed to incorporate this trait into commercially important cultivars.

There is a whole subspecies of hydrangeas presently unexplored by greenhouse forcers. The lacecaps (*H. macrophylla* subsp. *macrophylla* var. *normalis*) offer a more open inflorescence and more rapid forcing (Figure 7). These plants need to be examined for potential as a new florists' crop as well as a germplasm source for breeding programs. A 'Merritt's Supreme' type with variegated foliage shown in Figure 7 would be striking and may appeal to a new market to increase the diversity of hydrangeas.

Whatever the future holds for hydrangeas and other floricultural crops, the point needs to be made: Industry-supported applied research has advanced floriculture to standards unheard of 15 years ago, and recognition of that support is well-deserved and overdue. Too many questions need answers to stop now. I look forward to continued support and I am confident that new industry-supported research findings useful to hydrangea growers will be the cornerstone of the revised edition of this book.

References

1. Allen, R. C. 1931. Factors influencing the flower color of hydrangeas. *Proc. Amer. Soc. Hort. Sci.* 28:410.
2. Allen, R. C. 1934. Controlling the color of greenhouse hydrangeas (*Hydrangea macrophylla*) by soil treatments with aluminum sulfate and other materials. *Proc. Amer. Soc. Hort. Sci.* 32:632–634.
3. Allen, R. C. 1943. Influence of aluminum on the flower color of *Hydrangea macrophylla* DC. *Contr. Boyce Thompson Inst.* 13:221–242.
4. Allen, T. C. and W. C. Anderson. 1980. Production of virus-free ornamental plants in tissue culture. *Acta Hortic.* 110:245–251.
5. Anonymous. 1973. Good results achieved with growth regulators on hydrangeas. *Grower* (London) 80:647.
6. Arakawa, H. and S. Taga. 1969. Climate of Japan, p. 119–158. In: Arakawa, H. (ed.). *World Survey of Climatology, VIII, Climates of Northern and Eastern Asia.* Elsevier, New York.
7. Asen, S., H. W. Siegelman, and N. W. Stuart. 1957. Anthocyanin and other phenolic compounds in red and blue sepals of *Hydrangea macrophylla* var. Merveille. *Proc. Amer. Soc. Hort. Sci.* 69:561–569.
8. Asen, S., N. W. Stuart, and E. L. Cox. 1963. Sepal color of *Hydrangea macrophylla* as influenced by the source of nitrogen available to plants. *Proc. Amer. Soc. Hort. Sci.* 82:504–507.

9. Asen, S., N. W. Stuart, and A. W. Specht. 1960. Color of *Hydrangea macrophylla* sepals as influenced by the carry-over effects from summer applications of nitrogen, phosphorus, and potassium. *Proc. Amer. Soc. Hort. Sci.* 76:631–636.

10. Asen, S., N. W. Stuart, and H. W. Siegelman. 1959. Effect of various concentrations of nitrogen, phosphorus, and potassium on sepal color of *Hydrangea macrophylla. Proc. Amer. Soc. Hort. Sci.* 73:495–502.

11. Aylsworth, J. D. 1988. Biological controls offer promise for the future. *Greenhouse Grower* 6(3):28–32.

12. Bailey, D. A. 1983. *Control of growth and flowering in hydrangea* (Hydrangea macrophylla (Thunb.) Ser.). MS Thesis, Purdue Univ., West Lafayette, Ind.

13. Bailey, D. A. 1986. *Investigations of hydrangea distortion in florists' hydrangea* (Hydrangea macrophylla Thunb.). PhD Thesis, Purdue Univ., West Lafayette, Ind. (Diss. Abstr. 87-09769).

14. Bailey, D. A. 1989. Uniconazole effects on forcing of florists' hydrangeas. *HortScience* 24:In Press.

15. Bailey, D. A. and P. A. Hammer. 1988. Evaluation of nutrient deficiency and micronutrient toxicity symptoms in florists' hydrangea. *J. Amer. Soc. Hort. Sci.* 113: 363–367.

16. Bailey, D. A. and P. A. Hammer. 1989. Stimulation of "hydrangea distortion" through environmental manipulations. *J. Amer. Soc. Hort. Sci.* 114:411–416.

17. Bailey, D. A. and T. C. Weiler. 1984. Control of floral initiation in florists' hydrangea. *J. Amer. Soc. Hort. Sci.* 109:785–791.

18. Bailey, D. A. and T. C. Weiler. 1984. Stimulation of inflorescence expansion in florists' hydrangea. *J. Amer. Soc. Hort. Sci.* 109:792–794.

19. Bailey, D. A. and T. C. Weiler. 1984. Rapid propagation and establishment of florists' hydrangea.

HortScience 19:850–852.

20. Bailey, D. A., G. R. Seckinger, and P. A. Hammer. 1986. *In vitro* propagation of florists' hydrangea. *HortScience* 21:525–526.

21. Bailey, D. A., T. C. Weiler, and T. I. Kirk. 1986. Chemical stimulation of floral initiation in florists' hydrangea. *HortScience* 21:256–257.

22. Bing, A., J. W. Boodley, C. F. Gortzig, R. G. Helgesen, R. K. Horst, G. Johnson, R. W. Langhans, D. R. Price, J. G. Seeley, and C. E. Williamson. 1974. *Cornell Recommendations for Commercial Floriculture Crops, Parts I and II.* Cornell Univ. Agr. Bul., Ithaca, N.Y.

23. Dunham, C. W. 1948. *The culture and flowering of hydrangeas and azaleas as affected by growth habit.* MS Thesis, Univ. Wis., Madison.

24. Haworth-Booth, M. 1984. *The Hydrangeas.* 5th ed. Constable and Co., London.

25. Hara, H. 1955. Critical notes on some type specimens of East-Asiatic plants in foreign herbaria. *J. Japanese Bot.* 30(9):271–278.

26. Hunter, F. 1950. Hydrangea tests. *Ohio Florists' Assn. Bul.* 248:2–3.

27. Jones, J. B. 1979. Commercial use of tissue culture for the production of disease-free plants, p. 441–452. In: W. R. Sharp, P. O. Larsen, E. F. Paddock, and V. Raghaven (eds.). *Plant Cell and Tissue Culture.* Ohio State Univ. Press, Columbus, Ohio.

28. Jung, R. 1964. The status of hydrangea growing today. *Florists' Rev.* 135(3486):13–14, 35–37, 40.

29. Kenyon, O. 1972. Short term production of long lasting hydrangeas. *Minn. State Florists Bul.* Oct.:5–9.

30. Kiplinger, D. C. 1945. Well grown hydrangeas are valuable for the spring holidays. *Florists' Rev.* 95:23–25, 29–30.

31. Kofranek, A. M. and A. T. Leiser. 1958. Chemical defoliation of *Hydrangea macrophylla* Ser. *Proc. Amer. Soc. Hort. Sci.* 71:555–562.

32. Kohl, H. C., Jr. and R. L. Nelson. 1966. Controlling height of hydrangeas with growth retardants. *Calif. Agr.* 20(2):5.

33. Kosugi, K. and H. Arai. 1960. Studies on flower bud differentiation and development in some ornamental trees and shrubs. VII. On the date of flower bud differentiation and flower development in *Hydrangea macrophylla. Kagawa Daigaku Nogakubu Gakujutsu Hokoko* 12:78–83.

34. Koths, J. S., R. W. Judd, Jr., and J. J. Maisano, Jr. 1973. *Commercial Hydrangea Culture.* Univ. of Conn. Agr. Ext. Bul. 73–63.

35. Lawson, R. H. and R. K. Horst. 1983. Hydrangea diseases can be controlled. *Greenhouse Manager* 1(11):66–80.

36. LeMattre, P. 1975. Influence du facteur température sur la mis á fleur de l'Hortensia (*Hydrangea macrophylla*), p. 338–344. In: P. Chouard and N. de Bilderling (eds.). *Phytotronics in Agricultural and Horticultural Research III.* Gauthier-Villars, Paris.

37. Liberty Hyde Bailey Hortorium. 1976. *Hortus Third: a Concise Dictionary of Plants Cultivated in the United States and Canada.* 3rd ed. Macmillan, New York.

38. Link, C. B. and J. B. Shanks. 1952. Experiments on fertilizer levels for greenhouse hydrangeas. *Proc. Amer. Soc. Hort. Sci.* 60:449–458.

39. Litlere, B. and E. Strømme. 1975. The influence of temperature, daylength, and light intensity on flowering in *Hydrangea macrophylla* (Thunb.) Ser. *Acta Hortic.* 51:285–298.

40. Matsuzaka, Y. 1977. *Major Soil Groups in Japan.* Proc. Intl. Seminar on Soil Env. and Fertility Mgt. in Intensive Agr. pp. 89–95.

41. McClintock, E. 1957. A monograph of the genus Hydrangea. *Calif. Acad. Sci.* 29(5):147–256.

42. Miller, R. 1988. Making biological controls work for you. *GrowerTalks* 52(1):52–63.

43. Nakanishi, G., K. Yokoi, and S. Urabe. 1972. The studies on flower bud differentiation and flower bud development in *Hydrangea macrophylla* and the effect of low temperature treatment for breaking the dormancy on it. *Bul. of the Nara Agr. Expt. Sta.* 4:20–26.

44. Nelson, P. V. 1985. *Greenhouse Operation and Management.* Reston, Reston, Va.

45. Ohwi, J. 1965. *Flora of Japan.* Smithsonian Inst., Wash. D.C.

46. Papadakis, J. 1970. *Climates of the World: Their Classification, Similitudes, Differences and Geographic Distribution.* Buente, Buenos Aires.

47. Peters, J. 1975. Über die Blütenbildung einiger Sorten von *Hydrangea macrophylla. Gartenbauwissenschaft* 40(2):63–66.

48. Piringer, A. A. and N. W. Stuart. 1955. Responses of hydrangea to photoperiod. *Proc. Amer. Soc. Hort. Sci.* 65:446–454.

49. Piringer, A. A. and N. W. Stuart. 1958. Effects of supplemental light source and length of photoperiod on growth and flowering of hydrangea in the greenhouse. *Proc. Amer. Soc. Hort. Sci.* 71:579–584.

50. Pirone, P. P. 1978. *Diseases and Pests of Ornamental Plants.* Wiley and Sons, New York.

51. Post, K. 1942. Effects of daylength and temperature on growth and flowering of some florists' crops. *Cornell Univ Agr. Expt. Sta. Bul.* 787:46–48.

52. Post, K. 1959. *Florist Crop Production and Marketing.* Orange Judd, New York.

53. Powell, C. C. 1973. Botrytis blight of hydrangea. *Ohio Florists' Assn. Bul.* 528:3.

54. Ray, S. 1946. Reduction of blindness in hydrangeas. *Proc. Amer. Soc. Hort. Sci.* 47:501–502.

55. Robinson, G. M. 1939. Notes on variable colors of flower petals. *J. Amer. Chem. Soc.* 61:1606–1607.

56. Scott, B. 1982. Hydrangeas respond to new growth regulator. *N.C. Flower Growers' Bul.* 26:(4)10–12.

57. Shanks, J. B. 1969. Some effects and potential uses of ethrel on ornamental crops. *HortScience* 4:56–58.

58. Shanks, J. B. 1981. Out of season forcing of hydrangea. *HortScience* 16:83–84. (Abstr.).

59. Shanks, J. B. 1985. Hydrangeas, p. 535–558. In: V. Ball (ed.). *Ball Red Book.* 14th ed. Reston, Reston, Va.

60. Shanks, J. B. 1987. Development of ornamental crops under split night temperatures. *J. Amer. Soc. Hort. Sci.* 112:651–657.

61. Shanks, J. B., J. R. Haun, and C. B. Link. 1950. A preliminary study on the mineral nutrition of hydrangeas. *Proc. Amer. Soc. Hort. Sci.* 56:457–565.

62. Shanks, J. B. and C. B. Link. 1951. Effects of temperature and photoperiod on growth and flower formation in hydrangeas. *Proc. Amer. Soc. Hort. Sci.* 58:357–366.

63. Shanks, J. B., H. G. Mityga, and L. W. Douglass. 1986. Photoperiodic responses of hydrangea. *J. Amer. Soc. Hort. Sci.* 111:545–548.

64. Stolz, L. P. 1984. *In vitro* propagation and growth of hydrangea. *HortScience* 19:717–719.

65. Strain, J. R. 1984. How to figure your cost to grow bedding plants. *BPI News* 15(12):5–10.

66. Stuart, N. W. 1951. Greenhouse hydrangeas: some effects of artificial light, storage temperatures and fertilizers. *Florists' Rev.* 109(2813):37–40.

67. Stuart, N. W. and H. M. Cathey. 1962. Control of growth and flowering of *Chrysanthemum morifolium* and *Hydrangea macrophylla* by gibberellin. *Proc. Intl. Hort. Congr.* 15:391–399.

68. Stuart, N.W., A. A. Piringer, and H. A. Borthwick. 1955. Photoperiodic responses of hydrangeas. *Proc. Intl. Hort. Congr.* 14:337–341.

69. Swanson, C. L. 1946. Reconnaissance soil survey of Japan. *Soil Sci. Soc. Ann. Proc.* 11:493–507.

70. Takeda, K., M. Kariuda, and H. Itoi. 1985. Blueing of sepal colour of *Hydrangea macrophylla*. *Phytochemistry* 24:2251–2254.
71. Tayama, H. K. and V. Zrebiec. 1987. Growth regulator chart. *Ohio Florists' Assn. Bul.* 687:24–29.
72. Thomas, B. J., R. J. Barton, and A. Tuszynski. 1983. Hydrangea mosaic virus, a new ilarvirus from *Hydrangea macrophylla* (Saxifragaceae). *Ann. Appl. Biol.* 103:261–270.
73. Tjia, B. and J. Buxton. 1976. Influence of ethephon spray on defoliation and subsequent growth on *Hydrangea macrophylla* Thunb. *HortScience* 11:487–488.
74. Tjia, B., L. Stoltz, M. S. Sandhu, and J. Buxton. 1976. Surface active agent to increase effectiveness of surface penetration of ancymidol on hydrangea and Easter lily. *HortScience* 11:371–372.
75. Tracy, T. E. and A. J. Lewis. 1981. Effects of antitranspirants on hydrangea. *HortScience* 16:87–89.
76. Ulery, C. J. 1978. Quality hydrangea production. *Ohio Florists' Assn. Bul.* 582:3–4, 9.
77. U.S. Dept. of Agr. 1977. *Flowers and Foliage Plants.* Crop Rpt. Board Stat. Rpt. Serv. SpCr 6-1(77). Washington, D.C.
78. U.S. Dept. of Agr. 1978. *Floriculture Crops.* Crop Rpt. Board Econ., Stat., and Coop. Serv. SpCr 6-1(78). Washington, D.C.
79. U.S. Dept. of Agr. 1979. *Floriculture Crops.* Crop Rpt. Board Econ., Stat., and Coop. Serv. SpCr 6-1(79). Washington, D.C.
80. U.S. Dept. of Agr. 1980. *Floriculture Crops.* Crop Rpt. Board Econ., Stat., and Coop. Serv. SpCr 6-1(80). Washington, D.C.
81. U.S. Dept. of Agr. 1981. *Floriculture Crops.* Crop Rpt. Board Econ. and Stat. Serv. SpCr 6-1(81). Washington, D.C.
82. U.S. Dept. of Agr. 1982. *Floriculture Crops.* Crop Rpt.

Board Stat. Rpt. Serv. SpCr 6-1(82). Washington, D.C.

83. U.S. Dept. of Agr. 1985. *Floriculture Crops.* Crop Rpt. Board Stat. Rpt. Serv. SpCr 6-1(85). Washington, D.C.

84. U.S. Dept. of Agr. 1986. *Floriculture Crops.* Crop Rpt. Board Stat. Rpt. Serv. SpCr 6-1(86). Washington, D.C.

85. van Lenteren, J. C. and J. Woets. 1988. Biological and integrated pest control in greenhouses. *Ann. Rev. Entomol.* 33:239–269.

86. Weiler, T. C. 1980. Hydrangeas, p. 353–372. In: R. Larson, (ed.). *Introduction to Floriculture.* Academic Press, New York.

87. Weiler, T. C. and L. C. Lopes. 1974. Hydrangea distortion. *Focus on Floriculture* 2(2):9, Purdue Univ., West Lafayette, Ind.

88. Wilson, E. H. 1923. The hortensias. *J. Arnold Arb.* 4:233–246.

89. Wiśniewska, E. and Z. Zawadzka. 1962. The formation of inflorescence in *Hydrangea macrophylla* Ser. cv. Altona. *Acta Agrobotanica* 11:157–165.

90. Yock, N. 1988. Personal communication. Oregon Propagating Co., Brookings, Ore.

Appendix A

Useful conversions

°F	—	°C
30		−1.1
35		1.7
40		4.4
45		7.2
50		10.0
55		12.8
60		15.6
65		18.3
70		21.1
75		23.9
80		26.7
85		29.4
90		32.2

Inches	—	cm
1		2.54
2		5.08
3		7.62
4		10.16
5		12.70
6		15.24
7		17.78
8		20.32
9		22.86
10		25.40
11		27.94
12		30.48

Feet	—	meters
1		0.30
2		0.61
3		0.91
4		1.22
5		1.52
6		1.83

Ft-c —	Lux —	PPF $(\mu moles \cdot m^{-2} \cdot s^{-1})$[a]
1000	10,764	199
2000	21,571	399
3000	32,356	598
4000	43,142	797
5000	53,927	997
6000	64,712	1196
7000	75,498	1395
8000	86,283	1595
9000	97,069	1794
10,000	107,854	1993

[a]There is no direct conversion between foot-candle readings of light and readings of photosynthetic photon flux (between 400 and 700 rm). The conversion given is an approximation based on natural sunlight.

Gallons/200 ft² — ml/m²		Gallons	—	liters
1	204	0.25		0.946
		0.50		1.893
		0.75		2.839
		1.00		3.785
		2.00		7.571

Oz	—	grams	Oz/100 gal	—	mg/liter	Fl oz	—	ml
1		28.35	1		74.89	1		29.57
2		56.70	2		149.78	2		59.15
3		85.05	3		224.67	3		88.72
4		113.40	4		299.57	4		118.29
8		226.80	8		599.13	8		236.59
12		340.19	12		898.70	12		354.88
16		453.59	16		1198.26	16		473.18

Fl oz/100 gal — ml/100 liter	
1	7.8
2	15.6
3	23.4
4	31.3
8	62.5
12	93.8
16	125.0

Lbs	—	grams	Lbs/100 gal	— grams/liter
0.5		226.8	0.5	0.60
1.0		453.6	1.0	1.20
1.5		680.4	1.5	1.80
2.0		907.2	2.0	2.40
2.5		1134.0	2.5	3.00

Ozs/yd³ — grams/m³		Lbs/yd³ — grams/m³	
1	37.08	0.5	296.6
2	74.16	1.0	593.3
3	111.24	1.5	889.9
4	148.32	2.0	1186.6
8	296.64	2.5	1483.3

Appendix B

Desired characteristics of hydrangea growth media adapted from (22,44).

Character	Practical Use	Measurement	Recommended Range
Physical			
Dry Bulk Density	dry weight of media	dry weight per unit volume	10–45 lb/ft^3
Moist Bulk Density	weight of media during production; plant support	moist weight per unit volume; weigh at field capacity	40–75 lb/ft^3
Water Retention Capacity	media's ability to store water	% water by volume held after watering and drainage	35–60% (v/v)
Air Porosity	space available for air in media	% water by volume collected after watering to saturation	10–20% (v/v)
Total Pore Space	combined air and water pore volume	% by volume of water held when dry media in un-drained container is saturated with water	45–70% (v/v)

Character	Practical Use	Measurement	Recommended Range
Absorption Rate	ability of media to absorb water	inches of water absorbed per hour	> 2 in./hour
Chemical			
Cation Exchange Capacity	ability to store cations (nutrients)	submit sample to laboratory for analysis	by dry weight: 5–40 meq/100 g by volume 10–30 meq/100 cc
pH	acidity and alkalinity; affects avail- ability of nutrients	hydrogen ion concentration by pH meter	6.0–6.5 for pink flowers; 5.0–5.5 for blue flowers
Soluble Salts	salinity; ability of plants to take- up water; indi- cator of media nutrient status	electrical conductivity (EC) on an EC meter	0.5–1.5 dS/m for soil- containing and 1–2.2 dS/m for soilless media using 1 media: 2 water (by volume); 2–4 dS/m by paste extraction
C:N Ratio	indicator of N-deficiency situations	estimate using reference tables giving C:N ratios for media compo- nents and also the decom- position rates for the indi- vidual media components	≤ 30:1—no more than 30 lb C for every 1 lb of N

Character	Practical Use	Measurement	Recommended Range
Chorides	salt-causing ion; avoid high concentrations	soil and / or water analysis	<100 ppm by paste extraction
Sulfates	salt-causing ion; avoid high concentrations	soil and / or water analysis	< 1500 ppm by paste extraction
Available Nutrients	optimum fertility for plant growth and sepal coloration	depends on method used by analysis lab	see below

	Concentration in media extract (ppm)			
	Paste method[a]		Spurway method	
Nutrient	Pink sepals	Blue sepals	Pink sepals	Blue sepals
ammonium	<15	<15	<6	<6
nitrate	80–110	60–90	30–50	20–30
phosphorus	6–10	2–6	6–12	1–5
potassium	60–90	100–150	10–20	25–50
calcium	>100	>100	>100	>100

[a]Little work has been reported using paste extract for hydrangea nutrition monitoring, so the ppm given are intended as rough guidelines only.

The Politics of Space

The Politics of Space

A History of U.S.-Soviet/Russian Competition and Cooperation in Space

Matthew J. Von Bencke

WestviewPress

A Division of HarperCollins*Publishers*

Copyright © 1997 by Westview Press, A Division of HarperCollins Publishers, Inc.

Published in 1997 in the United States of America by Westview Press, 5500 Central Avenue, Boulder, Colorado 80301-2877, and in the United Kingdom by Westview Press, 12 Hid's Copse Road, Cumnor Hill, Oxford OX2 9JJ

Library of Congress Cataloging-in-Publication Data
Von Bencke, Matthew J.
   The politics of space : A history of U.S.-Soviet/Russian competition and
cooperation in space / Matthew J. Von Bencke.
      p.   cm.
   Includes bibliographical references and index.
   ISBN 0-8133-3192-7 (hardcover)
   1. Astronautics—International cooperation—Government policy—
United States.   2. Astronautics—International cooperation—
Government policy—Russia.   I. Title.
TL788.4.V65   1997
333.9'4—dc20   97-32067
                                                                                    CIP

10   9   8   7   6   5   4   3   2   1

Contents

Figures

Preface

My interest in the coexistence of U.S.-Soviet/Russian competition and cooperation in space grew out of my interests in American and Soviet government, international relations and space policy. The natural overlap of these disciplines is American and Soviet/Russian space policy and how space policy relates to foreign and domestic policies. This disciplinary overlap provides the basis for a history of the space age which is focused on U.S.-Soviet/Russian international space policy formation. Rooting this history in its broader contexts relates space to wider issues concerning domestic and foreign policy formation in both the United States and the Soviet Union/Russia.

The initial stage of this project was made possible by funding from the Harvard Russian Research Center, the Public Broadcasting Service (PBS) and the Ford Program. This funding facilitated research trips to Washington and Moscow. Unfortunately my first trip to Russia was cut short when I was mugged and needed to return to the West for medical treatment. The rampant crime in Russia is quite real and threatens would-be researchers as well as economic recovery! The Ford Program and PBS were especially helpful in enabling me to return as quickly as possible to Moscow, where I continued my interviews and archival research.

The final stage of this project was made possible by the Center for International Security and Arms Control (CISAC) at Stanford University. CISAC and its co-directors, David Holloway and Michael May, provided me with a wonderful environment for research and writing, and without their financial, intellectual and personal support I could not have completed this book.

Special thanks go to Viktor Sokol'sky, a long-time Russian space historian, who served as my host and mentor at the Russian Academy of Sciences' Institute of the History of Science and Technology, as well as to Leonid Vedeshin of Interkosmos and Vladimir Kurt of the Institute of Space Research, both of whom commented on various aspects of my work and introduced me to several other Russian specialists who did the same. I am similarly grateful for the hospitality and assistance of Roger Launius, Lee Saegesser and Dill Hunley of the NASA History Office.

Many people were kind enough to comment on various aspects of my work. For such constructive advice I thank Graham Allison, David Bernstein, Roger Bourke, Ashton Carter, A. Denis Clift, Timothy Colton, Richard Davies, Paul Doty, Igor Drovenikov, Lewis Franklin, Arnold Frutkin, Marshall Goldman, Alexander Gurshtein, David Holloway, Frank Holtzmann, Nicholas Johnson, Nikolai Kardeshov, John Logsdon, Michael May, Andrei Piontkowsky, Robert Putnam, Vladimir Semenov, Vyacheslav Slysh, Marcia Smith, Lt. General Bernard Trainor (U.S. Marine Corps, ret.), Adam Ulam, Daniel Usikov, Nikolai Vorontsov, Celeste Wallander, Frank Winter, Victor Zaslavsky and Charlie Zraket. However, they are not responsible for or necessarily in agreement with the views expressed here, nor are they to be blamed for any errors of fact or interpretation.

Lastly, I am greatly indebted to three individuals. First, Joseph Nye was especially helpful in guiding the formative stages of this research. Dr. Nye encouraged me to pursue my interests in astrophysics, the Soviet Union *and* foreign policy, and helped guide the beginning stages of what has become a delightful amalgam of intellectual pursuits.

Second, Loren Graham taught me a deep respect for the study of the history of science, helped me arrange my research in Russia and provided valuable commentary on drafts of my work. His kind assistance was a great encouragement and boon to me throughout this project.

Lastly, I owe countless thanks to Anthony Oettinger. Dr. Oettinger consistently provided invaluable counsel, essential perspective and, when needed and when deserved, warm encouragement. He tirelessly read dozens of iterations of this work, and, in the process, not only gave me terrific advice, but also taught me how to be a better writer and scholar. Indeed, his magnanimous nature pervades the entire Program on Information Resources Policy at Harvard University which he chairs. I give my deepest gratitude and respect to him.

Matthew J. Von Bencke

Introduction:
International Space Policy--
A Paradigm for
Intergovernmental Relations

[Y]ears had passed. The storm-tossed sea of...history had sunk to rest upon its shores. The sea appeared to be calm; but the mysterious forces that move humanity (mysterious because the laws that govern their action are unknown to us) were still at work.

-- L. N. Tolstoy in *War and Peace*[1]

Tolstoy wrote this about Europe after the Napoleonic Wars, but many people might argue that the same passage applies to the years following the end of the Cold War. In the early 1990s, after nearly fifty years of uneasy coexistence under a nuclear cloud, it finally seemed that Russia and the United States might work together, and that peaceful cooperation might become the order of the day. An especially tell-tale sign of the improved international climate was the beginnings of U.S.-Russian cooperation in space; after all, space had historically been a favorite realm of superpower competition. However, a close look at U.S.-Soviet/Newly Independent States interactions in space suggests that it would be unreasonable to bet too heavily on the end of competition in space (and in other areas, for that matter). After all, weren't the same "mysterious forces" governing humanity even after the Cold War ended?

"What is the future of the world's two biggest space programs?" is an important question policy makers, participants and observers began asking themselves in earnest after the end of the Cold War; by the middle of the 1990s the debate was far from over. Nor has the fate of the United States' and the former Soviet Union's space programs been decided. The future continues to hang in the balance. Should the U.S. and Newly Independent States (NIS) work together in space? If so, in what ways, and how? What are the risks involved in such cooperation, and how can we minimize them? In what ways will competition continue to play a role? How will the private sector influence all of these issues?

Moreover, the answers to all of these questions depends upon and at the same time influences the outcome of much larger issues: what is the future of American relations with the Newly Independent States?

The middle of the 1990s is the ideal time to examine these and other related questions. Sitting at this juncture in history not only can we look at the American and Soviet space programs; we also have collected enough post-Cold War experiences to begin to extrapolate further into the future. In addition, perhaps most importantly, now is the time when this future will be determined. This book is an attempt to digest the history of the interaction of the U.S. and Soviet/NIS space programs, Cold War and post-Cold War, to fill the gap left by Walter McDougall's 1985 *The Heavens and the Earth*.[2] It is also a political history of the space age which, unlike its predecessors, focuses on the intersections of the world's two biggest space programs and the larger contexts within which these programs exist. Such a compilation is needed to inform the policy makers, participants and observers of the world's space programs, i.e., those who will influence the years to come.

The best way to peer into the future is to examine history. A quick glimpse at three seminal events in the space age suggests that the answers to all of these questions are complex. First, for example, when the Soviets launched the first artificial satellite, Sputnik, as a part of the International Geophysical Year (IGY), the newspapers of what became the two great space powers did not say, "Soviets Launch First Satellite: International Scientific Community Gains." Instead, the *New York Times'* headline declared, "Device is Eight Times Heavier Than One Planned by U.S."[3] *Pravda* talked not of international harmony but proudly exclaimed that "our contemporaries will witness how the freed and conscientious labor of the people of the new socialist society makes the most daring dreams of mankind a reality."[4] Thus, although it emerged in the context of an international year promoting the exploitation of space for all mankind, Sputnik was the starting gun to a space race largely characterized by superpower competition.

Second, partly in response to Sputnik, the United States, in July 1958, created the National Aeronautics and Space Administration (NASA). NASA was designed to successfully prosecute the space race, i.e., to win the competition with the Soviets. However, at the same time, the National Aeronautics and Space Act of 1958 encouraged the new space Administration to pursue "a program of international cooperation."[5]

Lastly, in 1960, Senator Kennedy declared that, "We are in a strategic space race with the Russians, and we have been losing....we cannot run second in this vital race. To insure peace and freedom, we must be first."[6] In May 1961 President Kennedy decided that the U.S. would undertake the greatest space challenge it could hope to achieve--sending men to the moon and back within the decade. However, although Kennedy embraced and accelerated the space race, he also asked in a September 1961 address to the United Nations,

> Why, therefore, should man's first flight to the moon be a matter of national competition?...Surely we should explore whether the scientists and astronauts of the two countries--indeed of all the world--cannot work together in the conquest of space, sending some day in this decade to the moon not the representatives of a single nation, but the representative of all our countries.[7]

Why this consistent duality? Were the American and Soviet International Geophysical Year efforts tailored towards sharing knowledge with other space scientists, or were they tools of an image contest? Was NASA supposed to achieve U.S. leadership in space or forge bonds with other nations? Did Kennedy want to cooperate or compete in space?

The answer to all of these questions is various shades of "both." In fact, throughout the space age, there has been a coexistence of U.S.-Soviet/Russian competition and cooperation in space. Space policy is the confluence of a number of interests, both domestic and international. When the confluences of American and Soviet/Russian motivations in space have overlapped, the two space powers have collaborated; when their collections of interests have dictated competition, the United States and the Soviet Union/Russia have competed. The appropriate question is not, "Did the U.S. and the USSR/Russia compete or cooperate in space in such-and-such a year?" but "To what degree and in what ways did the U.S. and USSR/Russia compete and collaborate during such-and-such a period?" The complicated calculations involved with this analysis reflect the fact that the Soviet and American governments are each multifaceted entities with a myriad of coexisting, often competing, interests. American and Soviet/Russian space policy is an outgrowth of this fact, and the American and Soviet/Russian space programs' parallel pursuit of competitive and cooperative policies is in many ways a paradigm for intergovernmental relations. At its essence, this work sets forth how the United States and the Soviet Union/Russia have historically competed and cooperated in space.

There are three additional factors that have served to focus the study of this history. First, the subject of this study is rooted in the larger context of overall American and Soviet/Russian domestic and foreign policy concerns. Similarly, although this work concentrates on those aspects of the U.S. and Soviet/Russian space programs which are related to interaction with other countries' space activities, it relates these aspects with the countries' overall space programs.

Second, space has civilian and military applications and implications. An important aspect of both superpowers' space programs has always been the interplay between their civilian and military space goals. This study takes into account both sides of the states' space programs.[8] However, a full analysis of the countries' military space programs requires a great deal of information that remains, in accordance with the U.S. government's thirty-year rule, classified as of the mid-1990s. Thus, with the exception of the early space age (from the mid-1950s to the early 1960s) this work's analysis gives greater emphasis to the American and Soviet/Russian civilian space programs than to their military space programs. Nonetheless, Chapter One presents a model for understanding the relationship between civilian and military space programs which has continued throughout the space age; the Conclusion takes advantage of and expands upon this model, examining the two major space powers' post-Cold War military space programs.

Third, space programs can be broadly divided into piloted and unpiloted efforts. Although I discus both areas of both the American and Soviet/Russian space programs, I emphasize the manned projects slightly more. This decision stems largely from the facts that (1) the piloted aspects of civilian space programs consume the most resources, and, as such, tend to be accurate indicators of the whole within which they exist, and (2) manned missions tend to capture both the public's and top-level politicians' attention more than unmanned missions; thus, since this work focuses on political aspects of the space programs, piloted projects are more germane. An exhaustive account of the coexistence of competition and cooperation in space would spend more time discussing the unmanned space programs, but its findings would not differ significantly from the conclusions reached here.

Chapter One, "The Origins of Space Policy: A Reflection and an Instrument of Foreign and Domestic Interests," describes the birth and development of the space age after World War II through 1958, examining how both Eisenhower's and Khrushchev's space policies were a reflection and an instrument of their foreign and domestic policy goals. Eisenhower and Khrushchev turned to military applications of rocket technology hoping to bolster their militaries while spending less on defense. Khrushchev, for his part, kept the Soviet military and civilian space programs intertwined and exploited the Soviet space program to bolster the Soviet Union's, the Communist Party's and his own personal image. In contrast, Eisenhower tried to keep the American civilian and military space programs as separate as possible. In addition, facts which surfaced in the early 1990s help show that Eisenhower made a conscious decision not to race the Soviets to space. Instead he sought to give an appearance of promoting space for peace while simultaneously paving the way for the U.S. military space program.

Sputnik's effects on U.S. and world opinion began an acceleration of the space race. Chapters Two, "The Context of Competition and Cooperation in Space: Overall U.S.-Soviet Relations and the American and Soviet Space Programs," and Three, "Early Days of Confrontation and Attempts at Cooperation," collectively cover the period from 1957 to 1969. These chapters (1) trace the early space age competition, (2) show the importance of the larger contexts in which cooperation in space occurred, i.e., the nations' overall relations and space programs and (3) discuss U.S.-Soviet interaction in the United Nations Committee on the Peaceful Uses of Outer Space (COPUOS). Despite their Cold War animosity, as the world's two major space powers, the United States and the Soviet Union sometimes had overlapping interests in the COPUOS.

Chapter Four, "Cooperation Come and Gone: The 1975 Apollo-Soyuz Test Project and the Subsequent Souring of Superpower Relations," covers the period from 1969 to 1980, highlighting the first joint U.S.-Soviet manned mission in space. Although the Apollo-Soyuz Test Project was largely a show of image, it marked the first significant confluence of American and Soviet interests in space. This period also demonstrated that warm overall U.S.-Soviet relations were a prerequisite for collaboration in space: the ebb and flow of cooperation from 1969 to 1980 closely matched the onset and downfall of détente.

Chapter Five, "A Confluence of U.S. and Soviet/Russian Interests in Space: Cooperation as the Cold War Ends," covers the period from 1981-1995, examining how the end of the Cold War led to markedly improved superpower relations and, correspondingly, to unprecedented levels of U.S.-Soviet (and then U.S.-NIS) cooperation in space.

Chapters Six and Seven, "Obstacles to Cooperation in Space, Post-Cold War," take a second look at the late 1980s and early 1990s, showing that the Cold War did not end up all roses for collaboration in space. Chapter Six, "A Nation in Disarray," focuses on developments inside Russia, the main successor to the Soviet Union. In the early 1990s the Russian civilian space program struggled under potentially crippling circumstances: Russia often seemed to be flirting dangerously with economic, social and political catastrophe, and the former Soviet civil space and science communities were fighting to maintain their infrastructure, to overcome organizational difficulties and to deal with inherited and largely anachronistic conservatism.

In addition, the dissolution of the USSR left the Soviet space program spread among newly sovereign nations, nations whose troublesome relations introduced new problems for the former Soviet space program. Chapter Seven, "International Barriers, New and Old," examines international challenges to post-Cold War cooperation in space, including relations between Russia and Ukraine and Russia and Kazakhstan. It also takes another look at an old international issue: the relationship between launch vehicles and weapons technology.

As the space age matured and commercial satellite applications expanded, private companies began to play a larger role in space. These developments, combined with the opening up of the former Soviet Union's space programs, paved the way for U.S.-NIS commercial space ventures. Chapter Eight, "The Market Bridges the Gap: Commercial Space Cooperation," examines and categorizes a number of U.S.-NIS non- or quasi-governmental collaborative space ventures, evaluating their motivations and their relation to intergovernmental space cooperation. This chapter also analyzes some of the difficulties these commercial interactions faced.

The Conclusion, "The Space Age Outlives the Cold War," examines how the end of the Cold War affected U.S. and Russian space policy. A part of this examination is a close look at the future roles of both nations' military space programs. These and related developments suggest that, despite the important changes in the world, the basic rules which guide space interaction outlived the Cold War.

As Chapter Seven and the Conclusion detail, non-Russian former Soviet republics inherited important aspects of the Soviet Union's space program. Thus, non-Russian Newly Independent States, especially Ukraine and Kazakhstan, are discussed. However, Russia dominated the NIS's space activities. As a result, and in pursuit of readability, post-Soviet space activities are referred to as "Russian," despite the fact that they often involve other former Soviet republics.

Now, on to delineating these "mysterious forces...."

Notes

1. Trans. Rosemary Edmonds (London: Penguin Books, 1978) 1339.
2. New York: Basic Books, 1985.
3. 5 Oct. 1957: A1.
4. 5 Oct. 1957. As cited in F. K. Krieger, *Behind the Sputniks* (Washington: Public Affairs Press, 1958) 311-12.
5. "National Aeronautics and Space Act of 1958" (PL 85-568, 29 Jul. 1958), *United States Statutes at Large*, vol. 72, pt. 1, 432.
6. John F. Kennedy, "If the Soviets Control Space, They Can Control the Earth," *Missiles and Rockets* 10 Oct. 1960: 12.
7. *Public Papers of the Presidents of the United States: John F. Kennedy, 1963* (Washington: Government Printing Office, hereafter GPO, 1964) 695. Hereafter referred to as *PP of JFK*.

8. This is not to suggest that the nations' civilian and military space programs are independent of one another; indeed, NASA, the Department of Defense, the Central Intelligence Agency and the National Reconnaissance Organization cooperate extensively, and all Soviet/Russian launches have been conducted by the Soviet/Russian military. Thus, as Chapter One will further delineate, the civilian and military space programs of both countries overlap.

1

The Origins of Space Policy: A Reflection and an Instrument of Foreign and Domestic Interests

Impressions created will color the thinking of the entire Russian military establishment and set the stage for future relations. Every...man therefore has in his hands...a profound responsibility. Our performance will be the yardstick by which the Russians judge the fighting capabilities, the discipline, the morale and the energy of the whole of the American forces....

> -- indoctrination addressed to the 15th U.S. Army Air Force on the morning of its departure for Soviet air bases; this uneasy cooperative mission entailed the first major deployment of a U.S. military force on Soviet soil, June 1994[1]

In World War II, the United States and the Soviet Union had fought the Axis Powers together. At the war's end this uneasy cooperative relationship degenerated until the two former allies openly vied for influence around the world, fostering alliances which would project and protect their power. There were clearly two Great Powers with two diametrically opposed ideologies and the world polarized around them. The United States, after conflicts over Eastern Europe, Berlin, Greece, Iran and Turkey, adopted a policy of encircling the Soviet Union with allies, trying to contain the presumably expansionist Communist bloc. The Soviet Union established dominion over its Central and Eastern European buffer zone, expanded its influence in Africa, China and the rest of Asia and even made forays into the Western Hemisphere.

The post-war situation was fertile ground for such bipolar competition.

9

The Great Powers of the Concert of Europe were largely no more. Great Britain and France saw their power diminish greatly; Germany was divided and occupied. Meanwhile the United States had evolved from a largely isolationist state to one that dominated the global economy, the possessor of an unscathed homeland and an image as the world's savior. The U.S. military was well supplied, stationed around the world and had suffered, relative to the other major powers' armed forces, few casualties. The Soviet Union, for its part, had a massive army that, like its American counterpart, occupied lands far beyond its borders, a land mass second to none and a claim as the leader of a massive Sino-Soviet Communist bloc that preached worldwide revolution.

As it turned out, the post-World War II U.S.-Soviet competition remained largely a war of nerves, a war of images. The points of contention were outside the two Great Powers' own territories. Confrontations arose and tensions ran high, but neither side was willing to engage the other in hot wars. Thus was born the Cold War. In recruiting their allies, both superpowers sought to portray themselves as the side to be on, the inevitable winner, the owner of the true ideology. Each country tried to, as was later expressed, win the hearts and minds of the Third World.

Sputnik

On Friday, October 4, 1957, the Soviet Union successfully launched *Sputnik*, a 184-pound satellite. This "moon" was the first artificial satellite to orbit the earth. The dramatic evidence of Soviet scientific success shocked the world--especially the United States public.

Like a tide raised by the "red moon," a wave of fear rumbled across the Western world. The Soviets had seized the initiative and started the space race, jumping from the starting gate, while the United States had yet to take its place at the starting line. As Khrushchev himself bragged,

> we were the first to launch rockets into space...first, ahead of the United States, England, and France. Our accomplishments and our obvious might had a sobering effect on the aggressive forces in the United States....They knew that they had lost their chance to strike at us with impunity.[2]

The United States had come to depend on technology as its greatest advantage in the superpower competition. Its atomic monopoly had ended in 1949, but its drive to maintain technological superiority to offset its inferiority in conventional land forces had been enshrined in NSC-162/2, Eisenhower's "New Look" strategy. NSC-162/2, officially titled "Basic National Security Policy," cited the "great Soviet military power" and the "basic Soviet hostility to the non-communist world, particularly to the United States," and resolved "to meet the Soviet threat" without "seriously weakening the U.S. economy."[3] In pursuit of these two goals the New Look policy sought to increase America's ability to counter the Soviet military while cutting defense spending by 30% over four years. The Eisenhower administration, simply stated, sought "more bang for the buck." Rather than maintain a huge army (and, in the event of hostilities, face an unacceptable number of casualties), America would use its know-how, its advanced nuclear weapons, rockets, planes, tanks, surface ships, submarines, radar, etc., to counter the hundreds of thousands of Communist troops. Thus, by beating America in what was presumably a race into space, the Soviet Union seemed to threaten the foundation of the security of the United States and its allies.

Furthermore, this technological public relations victory suggested something much more fundamental. Achievements in space were the luxuries of an advanced state which was wealthy enough to push back the frontier of science. The satellite was a product of the intellectual, engineering and manufacturing elite of the country. In sum, a better space program was, it seemed, a reflection of a superior society. The U.S. House Select Committee on Astronautics and Space Exploration provided a voice to the American public's shock, disappointment and guilt, explaining that "much of the responsibility" for America's failure to be first in space "must rest with the people as a whole, who spurned what warnings there were of the evolving Soviet menace, until they could hear the radio beep or see the Soviet sputniks as they passed overhead."[4] Had America become lazy and allowed its Golden Age to end already? Was Soviet Communism producing a finer education system, smarter scientists, better research and development and more efficient manufacturing? After all, was not this technological triumph merely a reflection of the fact that Soviet Communism was superior to American capitalism?

The Soviet Union, the vanguard of the world socialist movement, certainly thought so. In launching Sputnik, the USSR was making a statement to the world about its superior system of government. As *Pravda* claimed in announcing Sputnik's launch,

Artificial earth satellites will pave the way to interplanetary travel and, apparently, our contemporaries will witness how the freed and conscientious labor of the people of the new socialist society makes the most daring dreams of mankind a reality.[5]

Seven years later in a similar article *Pravda* claimed that it was inevitable that the Soviets would be first in space: "This is natural...the so-called system of free enterprise is turning out to be powerless in competition with socialism in such a complex and modern area as space research."[6]

Space as a Reflection of Foreign and Domestic Policy

The Soviet Union

An examination of this, the dawn of the space age, reveals that for both superpowers the space program was an instrument and reflection of both domestic and foreign policy considerations. The interplay of these factors in the late 1950s and 1960s illustrates their coexistence throughout the space age.

The launching of Sputnik was a brilliant public relations coup, not a reflection of overall technological superiority.   Sputnik's main achievement, since it was launched on a long-range military missile, was the highly public demonstration that the USSR possessed an advanced ICBM program; it also allowed the Soviets to test their tracking, power and transmissions systems.   However, the satellite itself performed no specially designed experiments.   Khrushchev had pushed for the early launch in order to make sure that it was the Communist standard bearer which ushered in the Space Age, ostensibly promoting the peaceful exploitation of space for the good of all mankind. The timing was perfect for such a projection, since the launch took place during and was advertised as a part of the International Geophysical Year (IGY), a year dedicated to initiating the peaceful use of outer space for all nations' benefit.[7]

However, the main motivation behind Sputnik's program was not exploiting space for the good of the world--it was exploiting space to prove that Soviet communism was the world's best system of government. The victory improved the Soviet image around the world and built upon two other recent highly publicized Soviet successes: four years before Sputnik's flight the Soviets announced the explosion of what they claimed was the world's first deliverable hydrogen bomb, and six weeks before Sputnik's launch the Soviets claimed possession of the world's first ICBM.[8]   The Soviet space program had been ordered to launch the world's first satellite, and, as such, it was subject to foreign policy considerations.   Thus, the Soviet space program was largely a reflection and an instrument of Soviet foreign policy.

Khrushchev also hoped to achieve military parity with the United States. This would be no small task: the Soviet economy was smaller than the U.S.'s, and the USSR had started this race behind, exploding its first atomic bomb more than four years after the United States had already dropped two.   In addition, the Soviet population and infrastructure were heavily damaged during collectivization and the purges of the 1930s, not to mention during World War II, when the USSR suffered approximately 13,600,000 military and 7,720,000 civilian casualties (in contrast, the United States suffered 292,131 military and 5,662 civilian casualties).[9] By virtue of the fact that resources for the arms race came from the same budget which provided for domestic needs, this foreign policy goal was inextricably linked to Soviet domestic concerns.

One of the biggest problems Khrushchev faced in the mid-1950s was feeding the Soviet people. Soviet agriculture was stagnant--average farm yields did not increase from 1913 to 1953.[10] Famine loomed, and Khrushchev knew it. Compared to Western farming, Soviet agriculture was technologically backwards: Soviet farmers lacked the equipment, crop breeds, fertilizers and methods necessary to increase their efficiency. Self-proclaimed agricultural wizard Lysenko contributed to the problems by promoting outdated, untested and often backwards farming techniques. The looming disaster demanded rapid results, and this could only be achieved by dramatically increased investment.

However, this capital would not come easily. Khrushchev proposed to cut the military budget by one-third. In promoting this policy, Khrushchev embattled the huge Soviet military-industrial complex and all of its associated powers. The problem was, how could the Soviets compete successfully with American military power while spending less money on defense? How could Khrushchev placate the military-industrial complex?

The answer was a reliance on technology, on nuclear weapons. It was to these new weapons of mass destruction that Khrushchev, like his counterparts in the United States, turned for "more bang for the buck." The beauty of the logic must have seemed irresistible to a man in Khrushchev's position: a single nuclear weapon could wipe out an entire city or military base. Suddenly, instead of having to maintain enough divisions to be able to defeat the United States and its allies, one could argue that the Soviet Union needed only to build a nuclear arsenal capable of deterring, or, in the event of hostilities, destroying its enemies. A group of nuclear weapons would, in the long run, cost much less than the cost of maintaining a division and its support elements. Khrushchev openly acknowledged his reliance on nuclear weapons, stating in November 1959 that "[W]e now have stockpiled so many rockets, so many atomic and hydrogen warheads, that, if we were attacked, we would wipe from the face of the earth all of our probable opponents."[11]

All this meant more money for agriculture and the civilian economy. As Khrushchev reflected later in life,

Now that it's the size of our nuclear missile arsenal and not the size of our army that counts, I think the army should be reduced to an absolute minimum....When I led the Government and had final authority over the military allocations, our theoreticians calculated that we had the nuclear capacity to grind our enemies into dust, and since that time our nuclear capacity has been greatly intensified. During my leadership we accumulated enough weapons to destroy the principal cities of the United States, not to mention our potential enemies in Europe....If we try to compete with America in any but the most essential areas of military preparedness, we will be...exhausting our material resources without raising the living standard of our people. We must remember that the fewer people we have in the army, the more people we will have available for other, more productive kinds of work.... We must be prepared to strike back against our enemy, but we must also ask, 'Where is the end to this spiraling competition?'[12]

Still, developing the nuclear warheads and the missiles to deliver them was a gargantuan task. Khrushchev gave this research and development top priority; if he was to be able to save his country from agricultural disaster while competing with the United States militarily, the Soviet ICBM effort *had* to be successful. The research and development programs which were to launch Soviet satellites were the same as those which were to launch Soviet nuclear missiles; success in intercontinental rocketry was intimately tied with successes in space, and the increased emphasis on ICBMs served to accelerate the Soviet space program. In fact, Khrushchev allowed the satellite program to be born only as an offshoot of the military rocket program, once it had already proven successful.

Sputnik was launched on a modified *semyorka* missile, i.e., on the world's first ICBM (the NATO designation is "SS-6"). Khrushchev was hesitant to permit Korolev to launch a satellite because he feared it would delay the development of a deployable ICBM. Using a precious *semyorka* to launch a satellite would delay the missile program because a satellite · launch could not provide information about warhead electronics or reentry. The Soviet government had begun officially considering launching a satellite in May 1954; however, the program's fate rested on Khrushchev's personal approval. Korolev did not gain that approval until January 1956, when Khrushchev was visiting the missile center, Kaliningrad. At the end of a day discussing ICBMs, Khrushchev, as he was preparing to leave, asked Korolev if there was anything else he would like to discuss. Korolev responded by asking for permission to use a *semyorka* to launch a satellite. Khrushchev paused, and Korolev hurriedly interjected that "the Americans were already working on the launch of an artificial satellite." He claimed that the USSR could launch not only before the United States, but that it could launch a satellite several times bigger than the America's.

The opportunity to give the Americans a public "punch in the nose" enlivened Khrushchev, and he began asking Korolev how much the *semyorka* would have to be modified to launch a satellite. Korolev responded, "All we will do is remove the thermonuclear warhead and replace it with a satellite. That is all...."

Khrushchev responded, "If the main goal will not suffer, do it."[13] Many years later Khrushchev would remember this comment that became a seminal event in the history of the space age, writing,

> When we became convinced that Korolev had solved the problem of designing a rocket for space exploration, we were able to step into the international political arena and show that now even the territory of the United States of America was vulnerable to a strike by our missile forces. It became, as they say, a balm to the soul, and our position had improved.[14]

The Soviet Space Program and Khrushchev's Rise to Power

Khrushchev also used the space program to his own personal benefit. Khrushchev's position as the top Soviet leader was hard won and anything but stable. When Stalin died in March 1953, he left no clear successor. Although the Politburo declared active the principle of "collective leadership" in the wake of Stalin's death, Malenkov, Molotov, Kaganovich, Beria and Khrushchev all vied to increase their own personal power vis a vis the others.'[15] Few doubted that one leader would emerge preeminent in this system so tailored towards one-man rule; it simply remained to be seen who that person was.

Few people thought that man would be Nikita Khrushchev. In fact, in March 1953 the Politburo listed him as fifth in power.[16] Nonetheless, by 1957, after helping to subordinate the powerful state security apparatus to political leaders, taking advantage of the nomenklatura system to appoint loyal officials, arranging for the demotion of Malenkov and a reconciliation with Tito's Yugoslavia and delivering his remarkable anti-Stalin speech at the Twentieth Communist Party Congress in October 1956, Khrushchev stood at the pinnacle of Soviet power. However, this pinnacle was not stable; in June 1957 the Politburo demanded Khrushchev's resignation from his post as First Secretary, and it was only with the help of the military (as led by Marshal Zhukov) that Khrushchev survived this attack on his power. As a part of his struggle to power and of his effort to bolster his personal image as the Soviet Union's best-fit leader, Khrushchev made himself the space program's, and especially the cosmonauts,' personal patron.[17] In addition to directly intervening in rocketry research and development to make sure it received the resources it needed, Khrushchev arranged press conferences, parades and press accounts to portray himself as the genius behind Soviet space achievements. So it was, for example, with national fanfare that Khrushchev met and subsequently paraded Yuri Gagarin, the world's first man in space. It was only due to Khrushchev's personal interventions, Soviet citizens were to think, that the Soviets were so successful in space. This was a critical aspect of the creation of Khrushchev's public image-- indeed, of his own personality cult. He used this image to consolidate and increase his own power, or, in short, to improve his claim to being *the* leader of the Soviet Union. These were the Soviet Union's greatest public relations achievements, and it was Khrushchev who positioned himself to personally reap the image benefits.[18]

Khrushchev also inherited a greatly weakened Party. Stalin had risen above the Party and had worked around it. Under Stalin, Party Congresses were held extremely infrequently (for example, thirteen years elapsed between the Eighteenth and Nineteenth Congresses), the Politburo had been sub-divided (thereby increasing dissension and decreasing efficiency) and Party leaders had been arbitrarily removed, persecuted and executed. Leading politicians were subject to accusations, trials and punishments administered by a security apparatus subordinate only to Stalin. So weakened was the Party in 1953 that Malenkov chose to pursue the pinnacle of power by remaining Chairman of the Council of Ministers while yielding his position as First Party Secretary. It was Khrushchev who capitalized on this mistake, skillfully using his powers as First Secretary to his advantage.

In order to remain the Soviet Union's leader, Khrushchev had to strengthen the Party. By reinvigorating the Party, the First Secretary stood to gain in terms of personal power; indeed, for him there was no other way. Reinvigorating the Party also meant furthering (or, at the least, *appearing* to further) its ultimate victory, i.e., the advancement of the world socialist movement.

How could Khrushchev revitalize the Party and its revolutionary fervor? One way was to use propaganda, and space--that mysterious, dreamy, seemingly unconquerable void--was the perfect raw material for Khrushchev's propaganda mills. Khrushchev improved the Party's standing by declaring achievements in space to be the products of the Party-led socialist movement. Soviet successes in space, Khrushchev declared, were a product of the labor of all socialist people, a part of the advance of socialism towards achieving full knowledge and surpassing the doomed bourgeois imperialists.

Thus, the early Soviet space program was an instrument and reflection of both domestic and foreign policy goals. The Soviet space program was tailored largely to achieve dramatic image boosts. At the same time, in seeking to increase its global power reach by striving towards parity while avoiding an agricultural crisis, the USSR placed great emphasis on its rocket programs. Lastly, Khrushchev personally used the space program to bolster his and the Party's strength.

The United States

The relative openness of the American government facilitates a close-up examination of the formation of early U.S. space policy. American leaders, like their Soviet counterparts, sought to reduce their military budgets while increasing their military power largely through developing intercontinental ballistic missiles. However, unlike Khrushchev, Eisenhower resolved to keep the American military and civil rocket programs as separate as possible.

Rather than promote the most promising route for a quick satellite launch, the American leaders acted explicitly according to other, more diverse priorities. The United States, if its leaders had pushed for it, could have "beaten" Sputnik into orbit. The technology was there, but the priority was not. Keeping this in mind, early American space policy can be understood by analyzing three related priorities. First, its military missile programs (which were given top priority as a part of the New Look strategy) and civilian rocket development were kept as separate as possible. Second, in organizing the fledgling space program, the Eisenhower administration sought to project the U.S. as the leader in promoting space for peace. Third, the U.S. military and intelligence establishment was trying to lay the foundations of its future military space program. Investigating the manner in which the Eisenhower administration sought to realize these three priorities reveals that the early U.S. space program, like the Soviet program, was a product of foreign and domestic policy considerations.

What Race?

These non-race-oriented priorities reveal that the stark terms of the importance of image in the space race had, before Sputnik, not yet become clear to the American leadership. Though America's image turned out to be a driving force behind American's space program from its beginning, it is, ironically, that very image which suffered when it failed to launch first.

In 1957 America's leaders learned that not only was space to be an important part of the superpowers' international image competition, but that the space race would be, like the Cold War, total. There was no top-down push within the United States government to launch the world's first satellite. This decision resulted not so much from a dismissal of Soviet capabilities or from a confidence in America's "lead" as much as from a failure to recognize the value of launching a satellite first. President Eisenhower never pushed for the U.S. to be the first to do so; in fact, he insisted accurately five days after the Sputnik launch that the American "satellite program has never been conducted as a race with other nations."[19] Eisenhower, who played an active role in defense and related decisions, had specifically decided not to "race" the Soviet Union to space.

Reactions to Sputnik

It is useful to closely examine the context of Eisenhower's decision. The charged context of the Cold War competition highlights the fact that in deciding not to race the Soviet Union to space, the Eisenhower administration was placing higher priorities elsewhere. Moreover, this examination describes how with Sputnik the superpower image competition entered a new, more intense phase.

It is not that there were no predictions regarding the public relations impact the first satellite launch would have; in fact there were several such predictions which reached the highest levels of the executive branch. The stark and often perspicacious nature of the warnings highlights the importance the administration gave its other priorities.

First, to examine the warnings themselves. As early as 1946 a RAND study predicted that the first satellite launch "would inflame the imagination of mankind, and would probably produce repercussions in the world comparable to the explosion of the atomic bomb."[20] Similarly, NSC Directive 5520 warned that the Soviets were working on their own satellite program and noted that

> Considerable prestige and psychological benefits will accrue to the nation which first is successful in launching a satellite. The inference of such a demonstration of advanced technology and its unmistakable relationship to intercontinental ballistic missiles technology might have important repercussions on the political determination of free world countries to resist Communist threats, especially if the USSR were to be the first to establish a satellite.[21]

The top of the executive branch was aware of these potential repercussions. A classified report requested by President Truman and presented to President Eisenhower upon its completion predicted that a satellite

> would be considered of utmost value by the members of the Soviet Politbureau....the satellite would have the enormous advantage of influencing the minds of millions of people the world over during the so-called period

of 'cold war....' it should not be excluded that the Politbureau might like to take the *lead* in the development of a satellite. They may also decide to dispense with a lot of the complicated instrumentation that we would consider necessary to put into our satellite to accomplish the main purpose, namely, of putting a visible satellite into the heavens first. If the Soviet Union should accomplish this ahead of us it would be a serious blow to the technical and engineering prestige of America the world over. It would be used by Soviet propaganda for all it is worth.[22]

As the previous discussion of Sputnik's simplicity and Soviet propaganda shows, Dr. Grosse could hardly have been more correct. Also, in an April 27, 1955 letter to Dr. Alan Waterman, the Director of the National Science Foundation (which was overseeing the U.S. involvement in the IGY), Deputy Under Secretary of State Robert Murphy noted that the U.S. satellite project, if successful, would "undoubtedly add to the scientific prestige of the United States, and it would have a considerable propaganda value in the cold war."[23]

These predictions were largely borne out after Sputnik's launch. The world, entrenched in the Cold War and still gripped by the tension of the crises in Hungary, the Middle East and Poland, was carefully watching the superpower competition.[24] Domestic and world opinions were sensitive to what seemed to be the eclipsing of the leader of the West--in its supposed forte. American scientists dining with their Soviet counterparts in observation of the International Geophysical Year that October evening "were caught completely by surprise."[25] *The New Republic* declared that the launch was proof that "the Soviet Union has gained a commanding lead in certain vital sectors of the race for world scientific and technological supremacy,"[26] while the *New York Times* bemoaned the apparent relative feebleness of American plans with the headline "Device Is Eight Times Heavier Than One Planned by U.S."[27] *Newsweek,* in seeking to answer the "most chilling" question, "What effect would the Soviet achievement have on the nation's security?" noted that in the future "sputniks would be able to sight and even photograph just about every point on earth," and, most ominously, worried that the "successful launching of the satellite gave strong support to Russia's claim that it has a workable intercontinental ballistic missile."[28]

Politicians, reflecting and fueling the growing public hysteria, joined the chorus of the concerned. The White House was "engulfed in turmoil."[29] Senator Henry Jackson (D - Washington) called Sputnik "a devastating blow to the prestige of the United States as the leader in the scientific and technical world," the Democratic Advisory Council (which included Adlai Stevenson and former President Truman) concluded that "[t]he all-out effort of the Soviets to establish themselves as master of space [sic] around us must be met by all-out efforts of our own," and even President Eisenhower admitted that the Soviets had "gained a great psychological advantage through the world."[30]   Senator John Kennedy (D - Massachusetts), who would soon, as President, initiate the century's grandest space venture, emphasized the resulting global loss of prestige:

> It seems to me, and this is most dangerous for all of us, that we are in danger of losing the respect of the people of the world....[T]he people of the world respect achievement. For most of the 20th century they admired American science and American education, which was second to none. But now they are not at all certain about which way the future lies. The first vehicle in outer space was called sputnik, not Vanguard.[31]

Senator Lyndon Johnson (D - Texas) led those demanding to know if, how and why the United States had allowed its security to go lax and initiated the "Preparedness Investigating Subcommittee of the Committee on Armed Services." This subcommittee produced some 3290 pages of testimony, all because

> [o]ur country is disturbed over the tremendous military and scientific achievement of Russia. Our people have believed that in the field of scientific weapons and in technology and science that we were well ahead of Russia. With the launching of Sputniks I and II [*Sputnik II* was launched on November 3]...our supremacy and even our equality has been challenged.[32]

Indeed, from the public's point of view, American security and the technical supremacy upon which this security depended had been dramatically, convincingly challenged.[33] The Red nemesis' leader had declared "We will bury you," and the eery "beep, beep, beep" of Sputnik (as publicly broadcasted in the U.S.) echoed Khrushchev's threat all too loudly.

Fears that the United States was losing a scientific and technological race went beyond the White House meeting rooms. One national survey found that the "majority" of Americans agreed that "Sputnik means the Russians have beaten us scientifically."[34]   A congressional report concluded that "unless this country makes a larger effort, in the next ten to twenty years, it will find itself definitely left behind....the United States faces a very real possibility of being outdistanced on the scientific front in an age when science in application can decide the fate of nations."[35]

The American image suffered abroad, too. One survey showed that the majority of the citizens of New Delhi, Toronto, Paris, Oslo, Helsinki and Copenhagen agreed that Sputnik "struck a hard blow at U.S. prestige."[36] Such findings confirmed American leaders' fears that the disturbingly successful Soviet campaign to appear to be the world's scientific leader would scare allies into neutrality and neutral nations into the Eastern Bloc.

American politicians could not afford to stand by idly. Sputnik forced a firm response. After Sputnik it was beyond question that the space race was total. Not only did image dominate the space programs, making them instruments and reflections of foreign policy, but the resulting image was dominated by appearing to be the best--first. As the NASA Administrator who oversaw the Apollo era wrote, "The great issue of this age is whether the U.S. can, within the framework of the existing economic, social, and political institutions, organize its development and use of advanced technology as effectively for its goals as can the Soviet Union...."[37]

"Total" Space Race

Given this damage to the American image, why did the U.S. not try to launch its satellite as soon as possible? Part of the answer to this question lies in the advantage of hindsight. Eisenhower did not take the actions required to lead the United States into space first partly because he did not realize the public relations power of being first in space. Even after the Soviet scientific and public relations coup, Eisenhower declared, "I can't understand why the American people have got so worked up over this thing."[38] "Eisenhower was genuinely puzzled by the panic over Sputnik....[A]ccording to James Killian [Special Assistant to the President for Science and Technology], the President had no idea that the American public was 'so psychologically vulnerable.'"[39]

In addition, part of Eisenhower's inaction may be attributed to the overabundance of information and viewpoints and the pace with which the actual events unfolded. However, the most important factor to consider in understanding this crucial phase of the early American space program is that Eisenhower's decision to not push for as early a satellite launch as possible was mostly a reflection of different priorities.

Space for Peace

The significance of Eisenhower's decision not to prioritize as early a satellite launch as possible is highlighted by the fact that the United States, had speed been a priority, could have been the first nation to orbit an artificial satellite. The warnings and the technology existed, but the priorities were elsewhere. One of the priorities deemed more important than speed was the presentation of an image as promoting "space for peace," and this motivation helps explain why the different military space programs were handled the way they were.

Each military service had its own satellite program. The Army pursued and promoted its satellite program, which was run by the Army Ballistic Missile Agency (ABMA). Wernher von Braun, the leading German rocket scientist who fled, along with most of his colleagues, to the United States at the end of World War II, led the Army program. The ABMA proposed to use a modified Jupiter missile, the Jupiter-C, to launch a satellite. Meanwhile, the United States Air Force (USAF), asserting its dominion over what it defined as the "aerospace" continuum, presented its Thor and Atlas boosters as best suited to lead America into space (the Air Force was already in charge of the highly classified WS [Weapon System] 117L project, which designed and built America's first reconnaissance satellites; further discussion follows beginning on page 25). Lastly, the Navy had been researching satellites since 1945, and its Viking rocket was the centerpiece of its space program as run by the Naval Research Laboratory (NRL). Each of the three service's space organizations sought survival and growth at the expense of the others. The Army, Air Force and Navy--especially in the context of Eisenhower's budget cuts--each vied for the honor (and accompanying funding) which they assumed would come with the first satellite launch. Indeed, the survival of the ABMA, Air Force and NRL projects depended upon their ability to prove their unique worth.

These projects existed parallel to each other for several years, but, as the IGY drew closer, one of them had to be selected to provide the rocket for America's first satellite launch. Assistant Secretary of Defense for Research and Development Donald A. Quarles led the decision making process. Quarles initiated the advisory Stewart Committee (chaired by Homer Stewart of the Jet Propulsion Laboratory), which considered each service's proposal. The resulting decision came in a vote on August 3, 1955. Three committee members voted for the Navy proposal (Project Vanguard), two for the Army's (Project Orbiter).

This close vote, which determined that the Navy would be responsible for the United States' first satellite launch, illustrates the heated nature of the debate and at the same time reflects the administration's priorities. RAND and Quarles had told the committee members that the administration sought a space program with as strong a civilian character as possible. Eisenhower's public relations priority was to portray the American space program in terms of "space for peace." In terms of this criterion, the decision was an easy one: the Viking rocket was developed by private industry, whereas the Jupiter rocket was produced by an Army arsenal which was directed by von Braun, a former Nazi scientist whose V-2 rockets had damaged Allied targets during World War II. It is in view of these factors that Eisenhower decided "to entrust the admirals with the job."[40]

It is ironic that this public relations priority contributed to the decision which facilitated America's "losing" the satellite race, a loss which created a veritable public relations debacle. Vanguard not only failed to beat the Soviets into space, but it also failed to beat the ABMA--even though the Army project, after the Stewart Committee decision, was only marginally supported.[41] Clear evidence--much of which was available at the time of the decision--demonstrates that the ABMA could have launched a satellite sooner than the Navy.[42] In fact, recently declassified documents make it clear that the officials making the decision knew that, but simply did not prioritize speed. In revisiting the controversial decision the Department of Defense's Special Assistant for Guided Missiles, E. Murphree, wrote in a memorandum,

> I have looked further into the matter of the use of the Jupiter re-entry test vehicle as a possible satellite vehicle in order to obtain an earlier satellite capability as we discussed recently. I find that there is no question but that one attempt with a relatively small effort could be made in January 1957.[43]

In another memorandum Murphree wrote

> While it is true that the Vanguard group does not expect to make its first satellite attempt before August 1957, whereas a satellite attempt could be made by the Army Ballistic Missile Agency as early as January 1957, little would be gained by making such an early satellite attempt.[44]

Moreover, decision makers believed a January 1957 launch would not be as valuable as a later launch, both scientifically and politically:

> such a single flight would not fulfill the Nation's commitment for the International Geophysical Year because it would have to be made before the beginning of that period. Adequate tracking and observation equipment for the scientific utilization of results would not be available at this time.[45]

However, the administration's emphasis on a civilian program in pursuit of a positive image could prove effective only so long as the United States was the first to launch a satellite. The damage to America's image, domestically and internationally, which resulted from the Soviets beating the United States into space was, as the preceding descriptions of the hysteria and disillusionment indicate, greater than that which may have been caused by the near-subtlety of who produced the rocket which launched the world's first (*American*) satellite. The variation in military/civilian involvement between Project Vanguard and Project Orbiter was one of gradations: both projects were, in the end, controlled by the Navy and the Army, respectively.

Instead of stressing an early launch, the American satellite effort was emphasized as one aimed towards launching a satellite during the IGY which could provide useful data to scientists around the world. In a presidential statement of July 29, 1955 Eisenhower first announced that he had

> approved a plan by this country for going ahead with the launching of small unmanned Earth-circling satellites as part of the United States participation in the International Geophysical Year....the American program will provide scientists of all nations this important and unique opportunity for the advancement of science.[46]

As the President would point out after Sputnik's launch, the United States fulfilled these goals. From the start, the American space program emphasized long-term benefits from using satellites with advanced microelectronics in order to perform military and civilian functions including reconnaissance, communications, meteorology, navigation and measurements of the composition and properties of space. These priorities and America's strong microelectronics industry both resulted partly from and contributed to each other. *Explorer 1*'s tiny 10.5-pound package included micrometeoroid detectors, Geiger counters and telemetry equipment, and discovered the Van Allen radiation belts. All of the relevant data was presented openly to scientists through the IGY. This, the Americans fairly claimed, was the exploitation of space for the benefit of all mankind. Thus, the Vanguard civilian satellite program was favored to appear to be promoting space for peace and to keep the military and civilian rocket programs as separate as possible.[47]

Military Versus Civilian Rockets

There was, however, more at work than "space for peace." An October 9, 1957 White House statement accurately describes an aspect of America's early rocket programs' priorities, claiming that "Merging of this...effort [to launch an American scientific satellite] with military programs could have produced an orbiting United States satellite before now, but to the detriment of scientific goals and military progress." The desire to keep both the military and civilian rockets programs and to keep them apart was derived partly out of a desire to present the U.S. civilian space program as one pursuing space for peace, and partly out of the practical desire to keep the ICBM program on track:

> It would be impossible for the ABMA group to make any satellite attempt that has a reasonable chance of success without diversion of the efforts of their top-flight scientific personnel from the main course of the Jupiter

program, and to some extent, diversion of missiles from the early phase of the re-entry test program.

The United States ICBM program had, as of 1953, been roughly four years behind the Soviets'.[48] American security demanded that this gap be closed if not reversed; Eisenhower's hope to offset cuts in conventional forces with superior strategic forces magnified this need.

Space for Power: Paving the Way for American Eyes in the Sky

There was a third reason why the United States did not race to space. One of the primary uses American leaders saw for space was reconnaissance. The nature of the Cold War required military planners to prepare their forces to respond quickly to threats the world around. This required quick, effective, accurate intelligence—globally. This need was all the more urgent given that the Soviet Union was a closed, highly secretive society. The U.S. knew that the Soviets were developing new long-range bombers, overseas bases, submarine bases and perhaps most importantly ICBMs; the questions were where they were, when, and what were they doing. As Eisenhower stated his concern, "Modern weapons had made it easier for a hostile nation with a closed society to attack in secrecy and thus gain an advantage denied to the nation with an open society."[49]

Thus, even before Sputnik was launched, the United States gave a high priority to improving intelligence gathering and early warning capabilities. The American intelligence community was, in the early 1950s, desperate for information about the Soviet Union, and did its best with outdated German aerial reconnaissance photographs from the early 1940s, interrogations of returned German and Japanese prisoners of war and reconnaissance flights along Soviet borders. In pursuit of better sources of data the Air Force, with Eisenhower's approval, released 516 camera-equipped balloons from Western Europe in January and February 1956. However, the project was only marginally successful, as it depended on the balloons chancing to drift over useful, unobscured targets during the day.[50]

In order to improve upon the balloons the Air Force turned to high altitude piloted aircraft. The U-2 reconnaissance plane was built by Lockheed's California "skunk works" and could fly at over 70,000 feet (no fighter known of then could operate at more than 50,000 feet). In late 1955 the U-2 became, at Eisenhower's direction, primarily the Central Intelligence Agency's project,[51] and, in June 1956, the U-2 made its first flight over the Soviet Union. As was the case with Project Genetrix, the

U-2 overflights had Eisenhower's explicit approval. The U-2 was a dramatic improvement over the wandering balloons. Still, the strip of land a U-2 could cover was relatively limited. Furthermore, American leaders suspected that the Soviets would eventually develop effective countermeasures, and saw the U-2 as a stopgap used in anticipation of even more effective high altitude reconnaissance vehicles.[52]

The more effective vehicle they had in mind was the military satellite. Satellites could provide pictures more effectively than any means previously available, covering broad swaths of the earth extremely rapidly. Improvements in optics promised highly detailed images. Although satellites could not fly around clouds like planes often could, they were unstoppable in the sense that nothing could shoot them down. Furthermore, high-altitude reconnaissance could, in the event of hostilities, be an effective force multiplier, directing the appropriate forces to respond most efficiently at specific points; satellite meteorology could augment this multiplier.

Thus, American officials, beginning in studies of the mid-1940s, saw the space program as a future replacement for and improvement upon balloon and plane espionage. As early as 1945, interrogations of German World War II rocket scientists suggested the utility of space espionage.[53] An October 4, 1950 RAND study predicted that satellites could provide secure communications, "strategic and meteorological reconnaissance of the kind that...would be of high military value" and act as relay stations for guiding long-range missiles.[54] In March 1954 RAND predicted that a spy satellite system would produce 30 million pictures annually--a number equivalent to all the pictures the USAF had acquired from all sources in the previous twenty-five years,[55] and recommended that the "Air Force undertake the earliest possible completion and use of an efficient satellite reconnaissance vehicle" as a matter of "vital strategic interest to the United States."[56]

That same spring of 1954 Eisenhower, as a strategic thinker and former Supreme Allied Commander ever fearful of another Pearl Harbor-like surprise attack,[57] appointed a Surprise Attack Panel of leading scientists. The Surprise Attack Panel, which was renamed the Technological Capabilities Panel, had three subcommittees, one on offensive forces, one on defensive forces and one on intelligence, and was intended to propose ways to avoid surprise attack. James R. Killian, president of MIT from 1948 to 1959, chaired the Panel, and the intelligence subcommittee was led by Edward Purcell, a Nobel Prize-winning Harvard physicist, and Edwin Land, the inventor of the Polaroid camera, head of the Polaroid Corporation and a personal advisor to every American president from 1955 to 1970. On February 14, 1955 Killian and Land briefed Eisenhower on the Panel's final report, which concluded that

We must find ways to increase the number of hard facts upon which our intelligence estimates are based, to provide better strategic warning, to minimize surprise in the kind of attack, and to reduce the danger of gross overestimation or gross underestimation of the threat. To this end, we recommend the adoption of a vigorous program for the extensive use, in many intelligence procedures, of the most advanced knowledge in science and technology.

The form of this "most advanced knowledge" was omitted from the printed report for security reasons, but, as Killian and Land orally conveyed to the President, it was reconnaissance satellites.[58] Killian and Land urged Eisenhower to improve high altitude espionage capabilities, starting with balloons and airplanes and leading to satellites.[59] The historical development of Project Genetrix, the U-2 and Project Corona[60] show that Eisenhower followed their advice precisely.

Further research broadened the functions of military satellites. In the summer of 1958 NSC Directive 5814/1 predicted that military satellites would produce revolutionary advances in espionage, communications, weather forecasting, electronic countermeasures and navigation.[61]

Thus, from its beginning, the Air Force's 117L classified military space program received a high priority. The USAF asked RAND to study the issue, and in April 1951 RAND issued a report confirming the potential utility of spy satellites. RAND then subcontracted the issue to the Beacon Hill Study Group, which was chaired by Carl Overhage of Eastman Kodak. In June 1952 the Beacon Hill Study Group issued its final report, in which it concluded that improvements in electronics and optics could provide useful intelligence from high altitudes. In response the Air Force issued contracts to Radio Corporation of America, North American Aviation, Westinghouse Electric Corporation, Bendix Aviation, Allis-Chalmers and the Vitro Corporation to begin research and development of satellite television cameras, vehicle guidance equipment, attitude-control devices and nuclear auxiliary electrical power sources. On March 1, 1954 this private consortium, dubbed Project Feedback, concluded a common report detailing the hardware requirements for observation, cartographic and weather satellites and their ground support systems. The Air Force, riding the wave created by the Technology Capabilities Panel, awarded the first contract for spy satellite development to Lockheed on October 29, 1956, initiating what became Weapon System 117L. Project Vanguard had not even begun, but the road was thoroughly paved for America's military space program.[62]

However this road was not seen as without obstructions. One of the problems American planners saw with developing the necessary fleet of spy satellites was that the Soviets were sure to respond hostilely. Despite the American position that sovereignty over space did extend to space itself, the Soviets, they believed, were sure to object that American satellites were violating their sovereign air space.[63] In order to set a legal precedent for satellite overflight, the United States decided first to orbit as innocuous a satellite as possible. As NSC Directive 5520 said, "a small scientific satellite will provide a test of the principle of 'Freedom of Space.' The implications of this principle are being studied within the Executive Branch. However, preliminary studies indicate that there is no obstacle under international law to the launching of such a satellite."[64] A related RAND study titled, "The Satellite Rocket Vehicle: Political and Psychological Problems," concluded

> It is very doubtful, however, whether the USSR would accept such a 'vertical' limitation of sovereignty; and in any case, it would certainly not regard the passage of any vehicle in the outer space over its territory as 'innocent' if it were demonstrated that it was being used to perform acts which in themselves infringe upon sovereignty. We may assume that satellite operations designed to gather visual information in Soviet territory, if they become known to the Soviet leaders, will be construed by them as a 'consummated act of aggression.' This reaction is likely to verbalized in legalistic terms.[65]

In order to test the waters of what the Soviet leadership would consider "a major threat," the study concluded that "best policy seems to be to stress the experimental nature of instruments in, and communications with" a "preliminary" satellite. Then a "second 'work' satellite" "to be used for intelligence purposes" could be more safely launched.[66]

Early preparations for the satellite program show that this was in fact the course taken. The program, though intended to appear as civilian as possible, was initiated by Quarles at the Department of Defense.[67] Once under way, the International Geophysical Year satellite program was handled as a classified matter in close cooperation with the Department of Defense and the Central Intelligence Agency. In a classified May 1955 letter Dr. Waterman, the Director of the National Science Foundation, updated Quarles on the civilian project's progress. Dr. Waterman noted that he had "discussed the subject with Allen Dulles [Director, CIA], with Richard Bissell present, the latter being the one in Central Intelligence who is following this closely" and that "Dulles volunteered to present the subject to the Operations Coordinating Board in order to get action started." In addition, Waterman stated, the subject was "thoroughly canvassed" at the Department of State, and the Bureau of the Budget had "informally agreed as to the importance of the matter and their cooperation when action is needed. They agreed to keep the matter confidential...."[68] According to Waterman, the Department of State's approval was sought only after the plan had been drawn up with Quarles' guidance.[69] *Explorer 1* did more than introduce the United States to space, it specifically paved the way for the American military space program.

Meanwhile Quarles, who had become the Secretary of the Air Force, made sure that the military satellites came second, slowing down the Air Force's reconnaissance satellite construction in November 1956, less than a month after the USAF had issued the first development contract to Lockheed.[70] The military satellite would come, but its birth was carefully controlled to follow that of its civilian cousin.

Thus, the early American space program was arranged in such away as to present itself as promoting space for peace, to keep it as separate as possible from the development of ballistic missiles and to set a precedent for the military uses of space. The foreign policy considerations inherent in these international image and national security considerations were intertwined with Eisenhower's desire to derive more military might for less money.

America Awakened

The launch of Sputnik and the ensuing American reaction are seminal events in Cold War history. The shock of the Soviet launch, avoidable though it might have been, caused a heightened awareness among American leaders that space was crucial to the superpower image competition. Moreover, this Soviet propaganda victory elevated this image battle to a higher level, forcing the Americans to struggle to appear to "catch up." In this struggle the Eisenhower administration finally changed its original plans and allowed the Army to launch a satellite before Project Vanguard.

Not since Pearl Harbor had the American consciousness been so galvanized into action. Eight days after Sputnik's launch Eisenhower conferred with the Department of Defense's Office of Defense Mobilization's Science Advisory Committee (SAC) in what turned out to be a critical meeting. The SAC advisors (including Land and Killian) convinced Eisenhower that while the United States was still ahead of the Soviet Union in science and technology, it was rapidly losing this lead. Science was given a much higher priority in Russian culture, and their schools were a powerful reflection and guarantor of this fact. Unless the trend was reversed, Land argued, "Russian scientific culture will leave us behind as a decadent race."[71] Extrapolating from the current trends, the SAC concluded that in ten years the Soviet Union would be technologically superior. This meeting had a dramatic effect on

Eisenhower, who referred to it repeatedly in the weeks to come.[72]

Eisenhower and the Congress enacted broad changes in response to these warnings. These changes fit into three categories: education, governmental research and development organization and the American civil space program. First, in answer to the perceived crisis in education, on September 2, 1958 the President signed into law the National Defense Education Act (NDEA). Before this Act, American education had almost entirely been the responsibility of state and local governments. The NDEA established an unprecedented role for the federal government in education, authorizing just under $1 billion in primary and secondary education aid, a $295 million college loan fund which gave special consideration to students planning to teach or specialize in science, math or foreign languages, $280 million to help schools purchase laboratory equipment and an additional $59.4 million for graduate school students specializing in areas related to national defense.[73]

Second, Eisenhower reorganized the Department of Defense, strengthening the position of the Department's Secretary and trying to diminish the ill effects of interservice competition. Under the new arrangement the Secretary of Defense gained control of all operational commands and the entire Department of Defense budget, which he then dispersed to the individual services. The Joint Chiefs of Staff (JCS) was also strengthened and recast to be a cooperative body with an integrated staff designed to direct all armed forces in peace and in war. These measures were devised largely in order to increase the efficiency of research and development within the overall armed forces.

In addition, the Department of Defense's Science Advisory Committee was elevated to the White House, where it became the President's Science Advisory Committee (PSAC). Killian chaired this body as Eisenhower's first Special Assistant for Science and Technology from 1957 to 1959. This elevation was a part of the overall greater emphasis given to research and development, on which total government spending increased steadily from $4,462 million in 1957 to $4,991, $5,806 and $7,744 million in 1958, 1959 and 1960, respectively.[74]

Congress also acted to support and oversee future science and space policies. The Senate created the Committee on Aeronautical and Space Sciences and the House created the Committee on Science and Astronautics.

Lastly, on July 28, 1958, Eisenhower signed into law the National Aeronautics and Space Act of 1958, creating the civilian National Aeronautics and Space Administration (NASA).[75] NASA was expressly designed to make America the world's leader in space technology and exploration. The new Administration subsumed NACA (the National Advisory Committee for Aeronautics) and also absorbed Project

Vanguard, the Jet Propulsion Laboratory and part of the Army Ballistic Missile Agency. The civilian agency was to exercise "control over aeronautical and space activities sponsored by the United States, except [those]...primarily associated with the development of weapons systems, military operations, or the defense of the United States;" these responsibilities remained with the Department of Defense.[76]

With the new government involvement in education (especially scientific education), the streamlined federal research and development and Department of Defense structures and the new National Aeronautics and Space Administration Eisenhower formed the basis for America's participation in the newly begun space age. The United States would not stand by while the Soviets pressed their gains. As Eisenhower reflected in January 1958,

> what makes the Soviet threat unique in history is its all-inclusiveness. Every human activity is pressed into service as a weapon of expansion. Trade, economic development, military power, arts, science, education, the whole world of ideas--all are harnessed to this same chariot of expansion. The Soviets are, in short, waging *total cold war*.[77]

In the wake of Sputnik, America joined this battle, and the fight for world leadership entered a new realm in earnest.

Notes

1. Thomas A. Julian, "Operations at the Margin: Soviet Bases and Shuttle-Bombing," *The Journal of Military History* Oct. 1993: 640.

2. Nikita Sergeevich Khrushchev, *Khrushchev Remembers*, trans. Strobe Talbott (Boston: Little, Brown and Company, 1970) 516-17.

3. NSC 162/2 was the second version of report number 162 of the National Security Council (NSC), an executive advisory board which assists the president in the formation of national security policy. The National Security Council was created on July 25, 1947 as a part of the National Security Act. NSC 162/2 was classified until 1977, with access to it "very strictly limited on an absolute need-to-know basis," and was circulated to the NSC members, i.e., Secretary of the Treasury, Attorney General, Director of the Bureau of the Budget, Chairman of the Council of Economic Advisors, Chairman of the Atomic Energy Commission, the Federal Civil Defense Administrator, the Chairman of the Joint Chiefs of Staff and the Director of the Central Intelligence Agency. This report called for the development and maintenance of "a strong military posture, with emphasis on the capability of inflicting massive retaliatory damage by offensive striking power," an allied rapid deployment capability, a mobilization base "adequate to insure victory in the event of general war," a broad intelligence system and the maintenance of a "sound, strong and growing economy...[and] morale and free institutions." The report noted and predicted the continuance of U.S. and Soviet

reluctance to initiate general warfare; the atomic "stalemate" it described became the Cold War, and NSC 162/2 is largely the birth of American Cold War policy. President Eisenhower directed "all appropriate executive departments and agencies" to implement it immediately upon its dissemination October 30, 1953. Paul Kesaris, *Documents of the National Security Council 1947-1977* (Washington: University Publications of America, 1980) (reel 3, #1062, microfilm).

4. *The National Space Program* 85th Congress, 2nd Session (H. Rpt. 1758, *Series Set 12073*) (Washington: GPO, 1958) 30.

5. *Pravda* 5 Oct. 1957. As cited in F. K. Krieger, *Behind the Sputniks* (Washington: Public Affairs Press, 1958) 311-12.

6. *Pravda*, "Sorry, Apollo!" 11 Oct. 1964. Apollo was the American manned moon program; Chapter Three, pages 66-70. As cited in James E. Oberg, *Red Star in Orbit* (New York: Random House, 1981) 77.

7. The IGY was organized by the World Committee for the International Geophysical Year (CSAGI). In a 1954 Rome meeting CSAGI recommended that "thought be given to the launching of small satellite vehicles, to their scientific instrumentation, and to the new problems associated with satellite experiments, such as power supply, telemetering, and orientation of the vehicle." The IGY started on July 1, 1957 and ended on December 31, 1958.

8. The Soviet hydrogen bomb, called Joe-4, was exploded August 12, 1953, nine months after the United States exploded the world's first (non-deliverable) hydrogen bomb. Joe-4 (named after Joseph Stalin) was the product of a rapid catch-up effort and was the first sign that the Soviet Union was capable of competing with the United States technologically.

On August 27, 1957 the USSR announced possession of an ICBM after successful tests of a *semyorka* (R-7, or "ol' number seven"); this announcement was treated with a certain amount of skepticism by the United States public, but was given new credence by Sputnik's launch (page 19, note 28). The *semyorka* was designed chiefly by Sergei Korolev, the Soviet Union's leading rocket scientist.

9. Chris Cook and John Stevenson, *Longman Handbook of World History Since 1914* (New York: Longman, 1991) 50. The numbers for the Soviet Union are estimates.

10. Roy A. Medvedev and Zhores A. Medvedev, *Khrushchev: The Years in Power* (New York: W. W. Norton and Company, 1978): 27. Soviet crop yields were one-third those of other European countries in 1953.

11. Ervin Jerome Rokke, "Politics of Aerial Reconnaissance: Eisenhower Administration," diss., Harvard University, 1971, 125. Even as Khrushchev publicized his reliance on nuclear weapons, he greatly exaggerated his country's nuclear forces. Chapter Two, pages 45-46.

12. Khrushchev, *Khrushchev Remembers* 517-19.

13. Sergei Khrushchev, *Krizisi i Rakety* (Crises and Rockets), vol. 1 (Moscow: Novosti, 1994) 110-11. Sergei Khrushchev was the son of Nikita Khrushchev and accompanied his father on this and many other trips.

14. Nikita Sergeevich Khrushchev, *Khrushchev Remembers: The Glasnost Tapes* (Boston: Little, Brown and Company, 1990) 187.